BIOCHEMICAL SOCIETY SYMPOSIA

No. 78

RECENT ADVANCES IN MEMBRANE BIOCHEMISTRY

BIOCHEMICAL SOCIETY SYMPOSIUM No. 78
held at Robinson College, Cambridge, January 2011

Recent Advances in Membrane Biochemistry

ORGANIZED AND EDITED BY
J. MALCOLM EAST
AND FRANCESCO MICHELANGELI

Published by Portland Press Ltd on behalf of the Biochemical Society
Portland Press Limited
Third Floor, Charles Darwin House
12 Roger Street
London WC1N 4JU
U.K.
Tel.: +44 (0)20 7685 2410
Fax: +44 (0)20 7685 2469
Email: editorial@portlandpress.com
www.portlandpress.com

ISBN 978 1 85578 183 2
ISSN (print) 0067 8694

British Library Cataloguing-in-Publication Data
A catalogue record for this book is available from the British Library

Typeset by Aptara Inc., New Delhi, India
Printed in Great Britain by Bell & Bain Limited, Glasgow

Contents

Preface

When we began planning the Biochemical Society Annual Symposium on Recent Advances in Membrane Biochemistry back in May 2009, we had two aims. It had been some time since the biomembrane community in the U.K. had met *en masse* and a meeting was long overdue, and we also wanted to acknowledge Professor Tony Lee's contribution to the field of membrane biology in the U.K. on the eve of his retirement from the University of Southampton.

Figure 1. Professor Tony Lee with colleagues, friends and former laboratory members, at the Biochemical Society Annual Symposium dinner on 6 January 2011

We were keen that the Symposium should bring together a range of expertise across the spectrum from the more biological/physiological aspects of the field to the biophysical/structural elements of membrane biology. At the same time, we wanted to highlight the potential (some of it now realized) that this field has for human health. We were extremely fortunate to gather together many eminent international and national speakers for the 3 days.

The Symposium was subdivided into four sessions covering membrane protein expression and crystallization, the structure of membrane transporters, modelling of membrane protein structure, folding and dynamics, lipid–protein interactions (consequences for activity), and membrane proteins in pathology and disease. The intention was to provide an opportunity for lipid biochemists, protein crystallizers, protein expressers, protein modellers, biophysicists and biologists to get together and exchange ideas.

Many of the speakers at the Symposium have contributed to this volume of *Biochemical Society Symposia*[1], and the following chapters provide a reflection of the programme experienced by the delegates at the Annual Symposium.

Of course, we are all indebted to the protein expressers and crystallizers who have provided the biomembrane community with the biological material and the structural information to revolutionize the study of membrane proteins. Indeed, it was stated on more than one occasion during the Symposium that biomembrane research is currently undergoing a 'renaissance'.

To kick off the symposium, Roslyn Bill and her group at Aston University (see Chapter 1) provided insights into the biological 'bottlenecks' that impede the synthesis of heterologous membrane proteins by yeast cells. More importantly, perhaps, for the biomembrane community, they gave detailed information for optimizing growth conditions, and manipulating genes that limit protein synthesis, to effect up to 70-fold improvements in the yield of heterologously expressed membrane proteins.

The conventional method for the purification of membrane proteins involves prior solubilization of the membranes with detergent. One drawback to this approach is that the proteins are also separated from their lipid partners in the membrane. Tim Dafforn and his collaborators, who were selected to give an oral communication[2] based on the poster[3] that they presented at the Symposium, described a method involving the amphipathic polymer poly(styrene-*co*-maleic acid) that can reversibly encapsulate membrane proteins in a 10 nm disc-like structure that is free of the bilyer. These encapsulated proteins can then be purified along with a shell of native lipid, ensuring that the protein maintains its native activity [see *Biochem. Soc. Trans.* (2011) **39**, 813–818].

In recent years, there have been great strides in the elucidation of high-resolution structures of membrane proteins. At the end of 2010, 263 unique membrane protein structures had been determined (see the Stephen White Laboratory website, Membrane Proteins of Known 3D Structure, at http://blanco.biomol.uci.edu/Membrane_Proteins_xtal.html); however, most of these structures were of detergent-solubilized proteins. Martin Caffrey from Trinity College Dublin has outlined methodologies for crystallizing membrane proteins in a lipid environment through the use of lipidic mesophases and argues that the structures of proteins in such an environment are more likely to reflect the structures of proteins in their natural environment, the lipid bilayer (Chapter 2). This method has recently been used to produce a number of keenly anticipated structures detailing the conformational changes that accompany ligand binding to receptors.

Delegates at the Symposium were entertained and educated by presentations covering a wide range of high-resolution structures encompassing the whole

[1]The chapters in this volume are available online at http://symposia.biochemistry.org and have also been published in *Biochemical Society Transactions* volume 39, part 3 (http://www.biochemsoctrans.org).
[2]Articles based on these oral communications have been published in *Biochemical Society Transactions* volume 39, part 3 (http://www.biochemsoctrans.org).
[3]Abstracts submitted for all oral presentations and posters are hosted on the Biochemical Society's website (http://www.biochemistry.org).

gamut of membrane protein types, including channels, transporters and receptors, in both the poster sessions and the oral presentations. Ian Booth and his group provided a detailed description of the structure of the mechanosensitive channel MscS that is responsible for monitoring and responding to bilayer tension to ensure cellular integrity when bacteria are subjected to hypo-osmotic stress (Chapter 3). The relationship between the channel and its lipid environment and the mechanism by which lipid tension is coupled to gate opening were also examined. In contrast, Poul Nissen and his co-workers (Chapter 4) presented work covering a eukaryotic pump, the $\alpha 4$ catalytic subunit of the Na^+/K^+ pump from sperm. They outlined potential reasons as to why a unique isoform of the α-subunit should be required for this particular function/location and provided details of inhibitors that may be pivotal in developing male contraceptives.

We were also fortunate to have in this session the recipient of the Biochemical Society's Early Career Researcher Award, René Frank from the University of Cambridge, who provided an overview of the biology, structure and function of the ionotropic glutamate receptors and discussed the use of novel biochemical and genetic approaches for investigating ion channel receptor complexes [see *Biochem. Soc. Trans.* (2011) **39**, 707–718].

Despite the progress made in elucidating high-resolution membrane-protein structures by crystallographic techniques and other biophysical approaches, the vast array of data produced by the various genome projects means that modelling will continue to provide important insights into membrane protein structure for many proteins for the foreseeable future. Gunnar von Heijne and Paula Booth and her laboratory have approached this problem from opposite directions. von Heijne (Chapter 5) discussed how hydrophobicity scales can be generated from empirical observations of protein partitioning into lipid bilayers during co-translational protein insertion. He then discussed the development of algorithms based upon this scale and incorporating the empirically derived positive-inside rule; this was followed by an explanation of how marginally hydrophobic helices could be inserted into membranes in multispan membrane proteins. In contrast, Paula Booth (Chapter 6) investigated protein folding by following the unfolding of an intact membrane protein, the tetrameric prokaryote ABC (ATP-binding cassette) transporter BtuCD, during its denaturation with SDS or urea. The studies provide information about the subunit interactions, indicating that unfolding and disassembly of the subunits is a coupled process; these data provide a starting point for investigating complex assembly.

Since this Symposium was organized in part to acknowledge the contribution of the research of Tony Lee from the University of Southampton to our understanding of lipid–protein interactions, the section of the Symposium covering this aspect of research was particularly well supported. Tony Lee described lipid interactions with a range of membrane proteins: the Ca^{2+} pump; a mechanosensitive channel, MscL; and a K^+ channel, KcsA. He explained the importance of lipid bilayer thickness in matching the hydrophobic surface of membrane proteins and the binding of anionic annular lipids (i.e. those at the bulk lipid/protein interface) to particular binding positively charged 'hotspots' on certain proteins. In addition, the binding of anionic lipids to less-accessible non-annular sites was discussed (see Chapter 7). Richard Cogdell (Chapter 9) presented work from his laboratory that reveals the location of bound lipids and detergents in membrane protein crystal structures. Molecules of lipid and detergents modified with the heavy atom bromine can be identified in electron-density maps and he provided striking structures that vividly demonstrate the binding of lipids to the transmembranous domains of photosynthetic reaction centres. William Dowhan and his co-workers approach the question of the importance of lipid-protein interactions by quite a different route: they have prepared *Escherichia coli* mutant strains that through the lack of particular enzymes involved in lipid metabolism have modified membrane compositions. In particular. they lack PE (phosphatidylethanolamine), which leads to the incorrect insertion of the translocator, lactose permease (LacY), into the bacterial membrane and the loss of lactose uptake. This defect can be corrected by the addition of PE following LacY insertion, indicating that, independently of the translocon, protein-lipid interactions can directly affect membrane protein organization (see Chapter 8).

We wanted to emphasize, via the Symposium, the potential for human health that could result from a greater understanding of biomembrane structure and function. A highlight of this section of the Symposium was the presentation given by Professor Frances Ashcroft who eloquently outlined how an understanding of the function of ATP-sensitive K^+ channels (K_{ATP}) in pancreatic β-cells, neurons, muscle and endocrine cells has led to new therapies for patients with neonatal diabetes that result from mutations of K_{ATP} channels. In addition, Frank Wuytack and Peter Vangheluwe outlined the role of the cardiac SERCA2a (sarcoplasmic/endoplasmic reticulum Ca^{2+}-ATPase 2a) in modulating cardiac contraction and discussed potential strategies for modifying its activity to intervene in cardiac failure. The regulation of the SERCA2b isoform was also discussed, and the potential of a novel regulatory site present both on SERCA2a and SERCA2b was highlighted (Chapter 10). Two oral communications[4] based on posters[5] were presented as part of this section: Marc le Maire and his co-workers and collaborators also presented data relating to Ca^{2+} pumps and their potential as drug targets. They discuss the properties of PfATP6 (*Plasmodium falciparum* Ca^{2+}-ATPase). They have expressed PfATP6 in yeast and clearly demonstrated that this pump is not, as previously reported, inhibited by the anti-malarial artemisinin, but is inhibited by typical SERCA inhibitors such as thapsigargin and cyclopiazonic acid. These data indicate that PfATP6 is a potential target for anti-malarial drugs, but it is unlikely to be the site of action of the anti-malarial artemisinin [see *Biochem. Soc. Trans.* (2011) **39**, 823–831]. Ana Mata discussed data from her laboratory demonstrating that the neurotoxic amyloid β-peptide that accumulates in the brains of Alzheimer's dementia patients inhibits PMCAs (plasma membrane Ca^{2+}-ATPases). Since these pumps are pivotal for controlling cytoplasmic Ca^{2+} levels, their inhibition could lead to Ca^{2+} dysregulation and account for the neurotoxicity of the amyloid β-peptide

[4]Articles based on these oral communications have been published in *Biochemical Society Transactions* volume 39, part 3 (http://www.biochemsoctrans.org).

[5]Abstracts submitted for all oral presentations and posters are hosted on the Biochemical Society's website (http://www.biochemistry.org).

[see *Biochem. Soc. Trans.* (2011) **39**, 819–822]. Given the importance of SERCAs in modulating cellular activity, their potential as druggable targets and the finding that a number of toxic environmental pollutants inhibit these pumps, we have briefly reviewed the wide range of SERCA inhibitors. Their mechanisms of action and binding sites are explored and we discuss their potential clinical and pharmacological use (Chapter 11).

Judy Hirst and her group have an interest in human disease resulting from mutations to Complex I (NADH:ubiquinone oxidoreductase). Following from the recently published structure of bacterial complex I, they have re-examined data relating to sequence analysis, mutagenesis and mutations linked to human disease in the light of the structural data in an attempt to identify residues playing a major role in the structure, function and pathology of Complex I (see Chapter 12). In the closing session of the Symposium, Hendrick van Veen presented data from his group that provide information about the mechanism by which mutidrug transporters are able to pump anticancer agents and antibiotics from the cytoplasm of tumour cells and pathogenic bacteria respectively. Here they have outlined a model of the transport mechanism involving the dimerization of the ATP-binding sites leading to the exposure of a formerly inward facing drug-binding site to the external surface (see Chapter 13).

Looking back at the Symposium, it seemed that we had covered a lot of ground in a short space of time. This reflects the fact that the U.K. biomembrane community has not had a meeting of this type for some years and there was much to catch up on. With the availability of membrane protein structures continuing to increase and acting as catalysts for more directed investigations of structure and function the prospects for the future are very exciting. Perhaps an event such as this needs to be a more frequent fixture in the U.K. meetings calendar.

This preface is adapted from the Editors' introductory article to Biochemical Society Transactions volume 39, part 3 (pages 703–706).

J. Malcolm East
Francesco Michelangeli

Abbreviations

ABC	ATP-binding cassette
A-domain	actuator domain
AMP-PNP	adenosine 5′-[β,γ-imido]triphosphate
2APB	2-aminoethoxydiphenyl borate
$A_{2a}R$	A_{2a} adenosine receptor
BHQ	2,5-di-(t-butyl)-1,4-hydroquinone
2b-tail	SERCA2b C-terminus
CCPI	cyclic peptide inhibitor
CL	cardiolipin
CPA	cyclopiazonic acid
CRE	cAMP-responsive element
CREM	CRE modulator
CscB	sucrose permease
DDM	*n*-dodecyl β-D-maltopyranoside
DEER	double electron–electron resonance
DES	diethylstilbestrol
di($C_{14:1}$)PC	dimyristoleoylphosphatidylcholine
di($C_{18:1}$)PC	dioleoylphosphatidylcholine
di($C_{24:1}$)PC	dinervonylphosphatidylcholine
DoE	design-of-experiments
EGFP	enhanced green fluorescent protein
EM	electron microscopy
ER	endoplasmic reticulum
FRET	fluorescence resonance energy transfer
GabP	γ-aminobutyrate
GlcDAG	monoglucosyl diacylglycerol
GlcGlcDAG	diglucosyl diacylglycerol
GPCR	G-protein-coupled receptor
LacY	lactose permease
LDAO	*N*,*N*-dimethyldodecylamine-*N*-oxide
MAG	monoacylglycerol
MD	membrane domain
M-domain	transmembrane domain
MPDB	Membrane Protein Data Bank
MS&FB	Membrane Structural and Functional Biology
MX	macromolecular crystallography

NBD	nucleotide-binding domain
N-domain	nucleotide-binding domain
NHE	Na^+/H^+ exchanger
Ni-NTA	Ni^{2+}-nitrilotriacetate
OG	octyl β-D-glucoside
PC	phosphatidylcholine
P-domain	phosphorylation domain
PE	phosphatidylethanolamine
PG	phosphatidylglycerol
PheP	phenylalanine permease
PLB	phospholamban
PMCA	plasma membrane Ca^{2+}-ATPase
RC	reaction centre
SAXS	small-angle X-ray scattering
SBP	substrate-binding protein
SCAM™	substituted cysteine accessibility method as applied to TM orientation
SERCA	sarcoplasmic/endoplasmic reticulum Ca^{2+}-ATPase
SPCA	secretory pathway Ca^{2+}-ATPase
SR	sarcoplasmic reticulum
TBPPA	tetrabromobisphenol A
TFE	2,2,2-trifluoroethanol
TM	transmembrane (domain)
TMA-DPH	*N,N,N*-trimethyl-4-(6-phenyl-1,3,5-hexatrien-1-yl)phenylammonium *p*-toluenesulfonate
TMD	transmembrane domain
TMH	transmembrane helix
TMH4	transmembrane helix 4
TMRM	tetramethylrhodamine methyl ester
V_i	inorganic vanadate

Biochem. Soc. Symp. 78
Citation reference: Biochem. Soc. Trans. (2011) **39**, 719–723.

1

Understanding the yeast host cell response to recombinant membrane protein production

Zharain Bawa, Charlotte E. Bland, Nicklas Bonander, Nagamani Bora, Stephanie P. Cartwright, Michelle Clare, Matthew T. Conner, Richard A.J. Darby, Marvin V. Dilworth, William J. Holmes, Mohammed Jamshad, Sarah J. Routledge, Stephane R. Gross and Roslyn M. Bill[1]

School of Life and Health Sciences, Aston University, Aston Triangle, Birmingham B4 7ET, U.K.

Abstract

Membrane proteins are drug targets for a wide range of diseases. Having access to appropriate samples for further research underpins the pharmaceutical industry's strategy for developing new drugs. This is typically achieved by synthesizing a protein of interest in host cells that can be cultured on a large scale, allowing the isolation of the pure protein in quantities much higher than those found in the protein's native source. Yeast is a popular host as it is a eukaryote with similar synthetic machinery to that of the native human source cells of

[1]*To whom correspondence should be addressed (email r.m.bill@aston.ac.uk)*

many proteins of interest, while also being quick, easy and cheap to grow and process. Even in these cells, the production of human membrane proteins can be plagued by low functional yields; we wish to understand why. We have identified molecular mechanisms and culture parameters underpinning high yields and have consolidated our findings to engineer improved yeast host strains. By relieving the bottlenecks to recombinant membrane protein production in yeast, we aim to contribute to the drug discovery pipeline, while providing insight into translational processes.

Membrane proteins as targets for recombinant protein production

Membrane proteins are central to many cellular processes: they are involved in the uptake and export of diverse charged and uncharged molecules, as well as mediating the interaction of cells with their environment. As a consequence, they are of prime importance as drug targets to pharmaceutical companies. In order to structurally or functionally characterize a given member of this important class of protein, a sufficiently stable active sample is required. In practice, this means the requirement for a regular supply of milligram quantities of purified membrane protein [1].

Few membrane proteins have evolved to be naturally abundant in their native membranes with notable exceptions including mammalian and bacterial rhodopsins, aquaporins and complexes involved in respiration and photosynthesis. Inevitably, these proteins were among the first to have their structures solved: the first high-resolution crystallographic structure of a membrane protein [that of the photosynthetic reaction centre from *Blastochloris viridis* (formerly *Rhodopseudomonas viridis*) isolated from natural sources] was published in 1985 (referred to in [2]). The fact that it remained the sole unique structure of a membrane protein for the next decade is indicative of the inherent challenges of working with this class of protein. The first such challenge is that recombinant production is required in order to secure the hundreds of milligram quantities necessary to complete a successful structural biology project [3].

In 1998, the first structures derived from recombinant membrane proteins were deposited. These were of the prokaryotic proteins MscL [4] and KcsA [5], both produced in *Escherichia coli*. Thereafter, the elucidation of unique structures derived from recombinant sources saw an exponential-like growth rate, with 30 deposited in 2010 [1]. In contrast, the number of unique structures from native sources reached a plateau of around ten per year in 2002 [1]. Of particular note are eukaryotic membrane protein structures, which represent the most challenging targets: they account for fewer than 20 % of all the unique membrane protein structures deposited to date. The yeasts *Pichia pastoris* and *Saccharomyces cerevisiae* have made an important contribution to this count as they have been used to generate more than half of the recombinant eukaryotic membrane protein samples leading to structures, with the remainder coming from insect or mammalian cells [1]. Specific examples derived from recombinant yeasts include the aquaporins SoPiP2;1 [6] and AQP5 [7], the potassium channel K_v1.2 [8] and the calcium ATPase SERCA1a (sarcoplasmic/endoplasmic reticulum Ca^{2+}-ATPase 1a) [9].

Strategies for producing recombinant membrane proteins

The process of identifying a membrane protein target of interest to solving its structure at high resolution is complex, involving several potential bottlenecks [1]. The development of validated protocols and technologies for recombinant membrane protein production that are less reliant on traditional trial and error approaches is therefore key to progress in this area.

Unfortunately, producing recombinant membrane proteins can still seem like something of an artform. There is no guarantee of success when using a previously successful experimental set-up, even for targets that are highly homologous with those that worked well in earlier trials [10]. Suitable production hosts for this purpose include prokaryotic microbes such as bacteria, eukaryotic microbes such as yeast, higher eukaryotes such as insect and mammalian cells, and *in vitro* systems. All have their unique advantages and disadvantages [11]. We and others take the view that yeast provides a good compromise, benefitting from the speed, ease of use and low cost of goods of a microbe while having the benefits of a eukaryote. As we have already seen, this position is strengthened substantially by its increasingly important contribution to the structural biology of membrane proteins [1].

Our own work in this area was prompted by an inability to produce Fps1, a member of the aquaporin family, despite the fact that we could routinely produce orthologues such as GlpF [10]. We and others have therefore sought to understand the scientific principles that underpin how recombinant host cells can be used optimally for making membrane proteins. Some groups have chosen to tackle this problem by producing between 40 and 100 orthologues and then focusing on the protein that gives the best yields [12]. This approach has led to structures, but it is labour-intensive and by its nature does not necessarily yield the specific protein of interest. Our focus has therefore been to understand yeast cells in order to make them more amenable to producing the target of choice [13]. The strategy has been to increase the total volumetric yield of a production experiment, in particular by increasing the yield per cell or 'specific productivity'.

Improving volumetric yields in *S. cerevisiae* by metabolic engineering

One of the key features of the respiratory yeast species *P. pastoris* is that it can be grown to very high cell densities. Attenuances (D_{595}; widely referred to as 'OD' or 'optical densities') in the hundreds are achievable, which is a straightforward way of increasing volumetric yield. Conferring this property on respiro-fermentative *S. cerevisiae*, which sacrifices its biomass yield when producing ethanol, would be desirable as its genetics and molecular biology are much better established than those of *P. pastoris*. This strategy was therefore pursued by constructing a fully respiratory strain of *S. cerevisiae*: Otterstedt and colleagues replaced the endogenous hexose transporters in yeast with a single chimaeric transporter comprising the first six transmembrane domains of the low-affinity Hxt1 transporter and the last six transmembrane domains of the high-affinity Hxt7 transporter [14]. The resultant TM6* strain was found not to exhibit the respiro-fermentative behaviour of a wild-type yeast, but rather a monophasic metabolism consistent with being respiratory. This is most likely to be due to a restricted glucose consumption and hence a reduced glycolytic rate [14]. However, the important feature in this context is that TM6* was able to produce more protein than a wild-type strain by achieving a higher volumetric yield, albeit in the knowledge that its specific growth rate is approximately 67 % that of the wild-type parent [15].

In the TM6* strain, we found that the yield per cell of Fps1 was the same as that for two wild-type strains [15]. The increase in volumetric yield occurred, as anticipated, because the TM6* strain produces approximately 2–3-fold more biomass (determined by measurements of D_{595}) in culture than a wild-type strain. When it was used to produce two GPCRs (G-protein-coupled receptors), the A_{2a}R (A_{2a} adenosine receptor) and the cannabinoid receptor 2, the B_{max} and K_d values obtained were comparable with those from a wild-type strain. The increased volumetric yield was due to the fact that there was a doubling of D_{595} and, more relevantly, the total membrane protein yield in the TM6* strain (determined by protein assay) was 4-fold higher than in the wild-type strain [15].

Improving yields per cell of *S. cerevisiae* by strain engineering

Improving the biomass yield of a *S. cerevisiae* strain that also has an increased yield per cell could provide a route to maximal yields. We therefore looked at the global host cell response to producing a membrane protein [13]. Using arrays, we examined two different growth conditions that both led to relatively low protein yields (35°C, pH 5, and 35°C, pH7) compared with normal conditions (30°C, pH 5) and looked for changes in mRNAs that occurred in the same direction in both sets. This highlighted 39 genes which were validated against the list obtained when comparing the single growth condition that led to higher protein yields (20°C, pH 5) with normal conditions (30°C, pH 5). In all but one case, genes that were down-regulated under low-yielding conditions were up-regulated under high-yielding conditions and vice versa [16].

In order to determine whether these genes influenced the yield per cell, we first looked at all non-essential genes that were down-regulated under high-yielding conditions in a screen of deletion strains. This revealed, perhaps

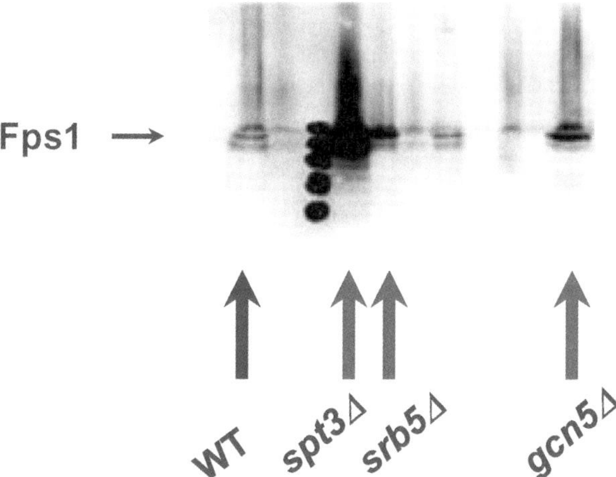

Figure 1. The deletion of a single gene can increase the yield of a recombinant membrane protein
The yield of a test protein, Fps1, was analysed from shake flask cultures of *S. cerevisiae* deletion strains using an anti-haemagglutinin immunoblot as the readout. Highlighted from the ten strains tested in this experiment (one per lane) are the yields from the wild-type (WT) parent strain and the three deletion strains, *srb5*Δ, *spt3*Δ and *gcn5*Δ, which clearly show yield improvements.

surprisingly, that, in some cases, the deletion of a single gene could increase the yield of our test protein compared with the wild-type strain, as seen for *SPT3*, *SRB5* and *GCN5* (Figure 1). The three deleted genes are known to be components of the transcriptional SAGA (for *GCN5* and *SPT3*) and Mediator (for *SRB5*) complexes. SAGA may have a role in the transcription of stress-induced genes in order to balance inducible stress responses with the steady output of housekeeping genes [17]. The Mediator complex appears to be required for all transcriptional events and transmits regulatory signals from transcription factors to RNA polymerase II [18]. We found that up to 9-fold improved yields of Fps1 over the corresponding wild-type control were not explained by changes in promoter activity or *FPS1* transcript number. This suggested that the improvements that we observed were not due to changes at the transcriptional level, but that post-transcriptional events might regulate the production of recombinant Fps1 [16].

Next, we looked at the overexpression of genes that were up-regulated under high-yielding conditions by using the doxycycline-repressible *tetO* system. This revealed an additional three candidate genes that might have a role in increasing the yield per cell of our membrane protein (Figure 2). Most notable among them was *BMS1*, which is involved in ribosome biogenesis. Using qRT-PCR (quantitative real-time PCR), we noted a clear relationship between Fps1 yield and an increase in *BMS1* transcript number compared with wild-type in all of our high-yielding host strains. In particular, we were able to tune the expression of

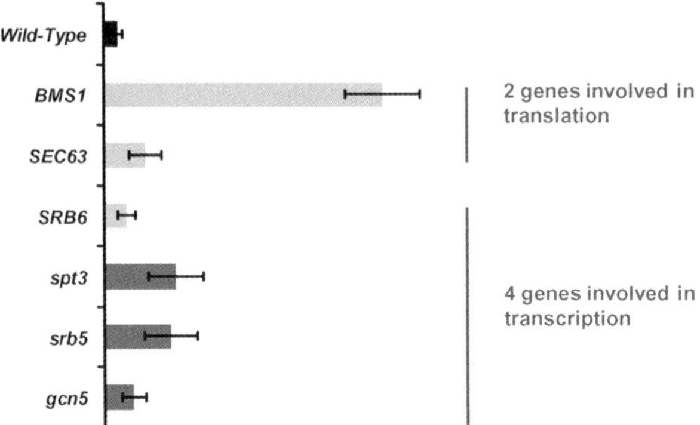

Figure 2. A selection of deletion and overexpression strains exhibiting improved yield characteristics for recombinant Fps1

Fps1 yields, as assessed by immunoblot analysis, are reported relative to wild-type (black bar). Lower-case letters denote deletion strains, whereas upper-case letters denote strains with an up-regulated expression of that particular gene.

BMS1 to maximize yields. When the promoter was fully on (with no doxycycline added) and the expression of *BMS1* was fully induced, we saw relatively low expression of our target protein. This was also the case when the promoter was fully repressed (at >10 μg/ml doxycycline). High-yielding strains and conditions were found to be correlated with 0.5–0.7 copies of *BMS1* mRNA per cell, with yield improvements of up to 70-fold. The use of this strain also doubled the yield of the GPCR $A_{2a}R$ after tuning with a different doxycycline concentration of 10 μg/ml [16].

Under high-yielding conditions, the cells grew more slowly and efficiently (indicated by a lower heat output rate as measured by online flow microcalorimetry) than under low-yielding or control conditions [16]. This is in agreement with the fact that these cells consume glucose and make ethanol more slowly, which may enable them to accommodate the increased metabolic load of producing higher yields of protein. This phenotype is also seen for our three high-yielding deletion strains. These strains also show differences in their translation characteristics and polysome profiles: they have significant accumulation of the polysomes into 80S peaks, strongly hinting at a block in initiation. This indicates changes in their translational properties compared with wild-type cells and links to an overarching role for *BMS1*. Selective translation of specific proteins is well documented following stress when ribosome functions are usually stalled at the initiation stage and where the Gcn2–Gcn4 complex plays a pivotal role [19–21]. It would therefore be interesting to establish whether such a regulatory complex is involved in the direct regulation of membrane protein production.

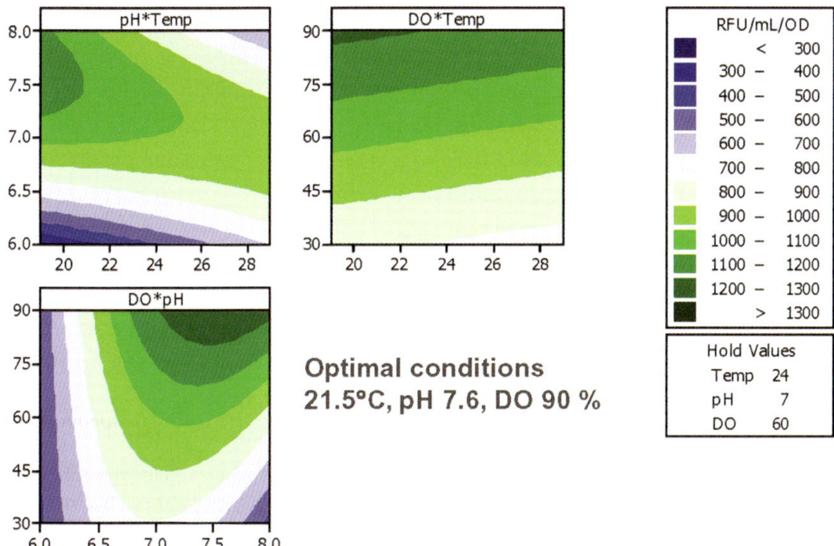

Figure 3. A response surface contour plot showing how yield per cell changes with each of the input factors

DO, dissolved oxygen tension (%); OD, attenuance units; RFU, relative fluorescence units; T, temperature (°C). All hold values are the '0' mid-point values in the DoE matrix.

Improving yields per cell of *P. pastoris* and *S. cerevisiae* by bioprocess control

In addition to genetic engineering, controlling culture parameters in both *P. pastoris* and *S. cerevisiae* has the potential to improve the yield per cell. Our initial work in *P. pastoris* was carried out on secreted GFP (green fluorescent protein), which is commonly used to tag membrane proteins [22]. Since we wanted to find a simple method of identifying and scaling up promising production conditions (thereby minimizing trial and error) we adopted a DoE (design-of-experiments) approach, which is often used in industry, to examine the effect of three input parameters on the product yield. The DoE approach has several advantages since it reduces the number of experiments that need to be performed to examine the combined effect of the input parameters, allows any interactions between them to be interrogated and results in a predictive equation.

We used a Box–Behnken model for our DoE, with three input parameters (temperature, pH and dissolved oxygen) set at three levels. This statistical design requires only a subset of all possible combinations to be used to identify the optimal production condition; in this case, 13 experiments out of a possible 27. Our readouts were the D_{595} of the culture and its fluorescence. We therefore reported the yield per cell as the fluorescence per D_{595}. Experiments were performed in a parallel mini-bioreactor system that can individually control pH, temperature and dissolved oxygen in each of its 24 vessels. This approach allowed us to develop a model that predicted the optimal yielding conditions for our culture (Figure 3), which we validated by comparing predicted with experimental yields for additional combinations of factors not used to construct the model. We demonstrated further that the model was scalable from 6 ml to 3 litres, although the relationship here suggested that it was under-predicting the yield. This led us to explore the induction regime, which is a critical component of a *P. pastoris* production experiment [22].

We noted that induction with 100 % methanol did not result in a sustained production, and that, after 15 h, the total yield had typically reached a plateau. We therefore examined a mixed induction of 60 % sorbitol and 40 % methanol, and saw that production was sustained, and, in the case of our DoE-optimized conditions, the yield was dramatically improved. Gas chromatographic analysis of the residual methanol concentration showed that the best yielding conditions had residual methanol concentrations in the region of 3 g/l. At concentrations below this, yields were lower. This optimized feeding regime increased the yield per cell approximately 6-fold over standard conditions.

We have recently applied the DoE approach to the production of a GPCR in three strains of *S. cerevisiae* using three culture parameters as before. In addition, the design includes the absence or presence of additives that have been shown to improve the functional yields of some GPCRs. As predicted, the input parameters did not significantly affect the growth of any of the three strains. However, they had a clear effect on binding activity in the membranes, with each strain giving its highest values under different culture conditions. The complete dataset will be used to build a predictive model, which will then be validated at scale.

Conclusions

Yeast is a flexible recombinant host for eukaryotic membrane protein production. The conditions under which yeast cells are grown and harvested are critical in maximizing the total volumetric yield of functional protein. Moreover, differences in yields can be related to the differential expression of genes involved in transcription and translation, and these genes have provided the basis for developing new production strains that can improve yields per cell up to 70-fold. We have used these strategies to produce several membrane proteins in yields suitable for further study. Examples include three four-transmembrane domain proteins (CD81, CD82 and CLDN1), our initial target, Fps1, and the GPCR $A_{2a}R$.

Funding

This work was supported by the European Commission [via contracts LSHG-CT-2006–037793 (OptiCryst) and HEALTH-F4–2007-201924 (EDICT: European Drug Initiative on

Channels and Transporters)], Engineering and Physical Sciences Research Council [via a CASE (Co-operative Awards in Science and Engineering) studentship with Applikon Biotechnology] and Biotechnology and Biological Sciences Research Council (via CASE studentships with AstraZeneca Ltd and Glycoform Ltd, as well as a Targeted Priority Studentship in Ageing) to R.M.B.

References

1. Bill, R.M., Henderson, P.J.F., Iwata, S., Kunji, E.R.S., Michel, H., Neutze, R., Newstead, S., Poolman, B., Tate, C.G. & Vogel, H. (2011) Overcoming barriers to membrane protein structure determination. *Nat. Biotechnol.* **29**, 335–340
2. Deisenhofer, J., Epp, O., Sinning, I. & Michel, H. (1995) Crystallographic refinement at 2.3 Å resolution and refined model of the photosynthetic reaction centre from *Rhodopseudomonas viridis*. *J. Mol. Biol.* **246**, 429–457
3. Bonander, N. & Bill, R.M. (2009) Relieving the first bottleneck in the drug discovery pipeline: using array technologies to rationalize membrane protein production. Expert Rev. *Proteomics* **6**, 501–505
4. Chang, G., Spencer, R.H., Lee, A.T., Barclay, M.T. & Rees, D.C. (1998) Structure of the MscL homolog from *Mycobacterium tuberculosis*: a gated mechanosensitive ion channel. *Science* **282**, 2220–2226
5. Doyle, D.A., Morais Cabral, J., Pfuetzner, R.A., Kuo, A., Gulbis, J.M., Cohen, S.L., Chait, B.T. & MacKinnon, R. (1998) The structure of the potassium channel: molecular basis of K^+ conduction and selectivity. *Science* **280**, 69–77
6. Tornroth-Horsefield, S., Wang, Y., Hedfalk, K., Johanson, U., Karlsson, M., Tajkhorshid, E., Neutze, R. & Kjellbom, P. (2006) Structural mechanism of plant aquaporin gating. *Nature* **439**, 688–694
7. Horsefield, R., Norden, K., Fellert, M., Backmark, A., Tornroth-Horsefield, S., Terwisscha van Scheltinga, A.C., Kvassman, J., Kjellbom, P., Johanson, U. & Neutze, R. (2008) High-resolution X-ray structure of human aquaporin 5. *Proc. Natl. Acad. Sci. U.S.A.* **105**, 13327–13332
8. Long, S.B., Campbell, E.B. & Mackinnon, R. (2005) Crystal structure of a mammalian voltage-dependent *Shaker* family K^+ channel. *Science* **309**, 897–903
9. Jidenko, M., Nielsen, R.C., Sorensen, T.L., Moller, J.V., le Maire, M., Nissen, P. & Jaxel, C. (2005) Crystallization of a mammalian membrane protein overexpressed in *Saccharomyces cerevisiae*. *Proc. Natl. Acad. Sci. U.S.A.* **102**, 11687–11691
10. Hedfalk, K., Bill, R.M., Hohmann, S. & Rydström, J. (2000) Overexpression and purification of the glycerol transport facilitators, Fps1 and GlpF, in *Saccharomyces cerevisiae* and *Escherichia coli*. in *Molecular Biology and Physiology of Water and Solute Transport* (Hohman, S. & Nielsen, S., eds.), pp. 29–34, Kluwer Academic/Plenum Publishers, Amsterdam
11. Grisshammer, R. (2006) Understanding recombinant expression of membrane proteins. *Curr. Opin. Biotechnol.* **17**, 337–340
12. Jiang, Y., Lee, A., Chen, J., Ruta, V., Cadene, M., Chait, B.T. & MacKinnon, R. (2003) X-ray structure of a voltage-dependent K^+ channel. *Nature* **423**, 33–41
13. Bonander, N., Hedfalk, K., Larsson, C., Mostad, P., Chang, C., Gustafsson, L. & Bill, R.M. (2005) Design of improved membrane protein production experiments: quantitation of the host response. *Protein Sci.* **14**, 1729–1740
14. Otterstedt, K., Larsson, C., Bill, R.M., Stahlberg, A., Boles, E., Hohmann, S. & Gustafsson, L. (2004) Switching the mode of metabolism in the yeast *Saccharomyces cerevisiae*. *EMBO Rep* **5**, 532–537
15. Ferndahl, C., Bonander, N., Logez, C., Wagner, R., Gustafsson, L., Larsson, C., Hedfalk, K., Darby, R.A. & Bill, R.M. (2010) Increasing cell biomass in *Saccharomyces cerevisiae* increases recombinant protein yield: the use of a respiratory strain as a microbial cell factory. *Microb. Cell Fact.* **9**, 47

16. Bonander, N., Darby, R.A., Grgic, L., Bora, N., Wen, J., Brogna, S., Poyner, D.R., O'Neill, M.A. & Bill, R.M. (2009) Altering the ribosomal subunit ratio in yeast maximizes recombinant protein yield. *Microb. Cell Fact.* **8**, 10
17. Huisinga, K.L. & Pugh, B.F. (2004) A genome-wide housekeeping role for TFIID and a highly regulated stress-related role for SAGA in *Saccharomyces cerevisiae*. *Mol. Cell* **13**, 573–585
18. Kornberg, R.D. (2005) Mediator and the mechanism of transcriptional activation. *Trends Biochem. Sci.* **30**, 235–239
19. Yang, R., Wek, S.A. & Wek, R.C. (2000) Glucose limitation induces GCN4 translation by activation of Gcn2 protein kinase. *Mol. Cell. Biol.* **20**, 2706–2717
20. Natarajan, K., Meyer, M.R., Jackson, B.M., Slade, D., Roberts, C., Hinnebusch, A.G. & Marton, M.J. (2001) Transcriptional profiling shows that Gcn4p is a master regulator of gene expression during amino acid starvation in yeast. *Mol. Cell. Biol.* **21**, 4347–4368
21. Valenzuela, L., Aranda, C. & Gonzalez, A. (2001) TOR modulates GCN4-dependent expression of genes turned on by nitrogen limitation. *J. Bacteriol.* **183**, 2331–2334
22. Holmes, W.J., Darby, R.A., Wilks, M.D., Smith, R. & Bill, R.M. (2009) Developing a scalable model of recombinant protein yield from *Pichia pastoris*: the influence of culture conditions, biomass and induction regime. *Microb. Cell Fact.* **8**, 35

Biochem. Soc. Symp. 78
Citation reference: Biochem. Soc. Trans. (2011) **39**, 725–732.

2

Crystallizing membrane proteins for structure–function studies using lipidic mesophases

Martin Caffrey[1]

Membrane Structural and Functional Biology Group, School of Biochemistry and Immunology, and School of Medicine, Trinity College Dublin, Dublin 2, Ireland

Abstract

The lipidic cubic phase method for crystallizing membrane proteins has posted some high-profile successes recently. This is especially true in the area of G-protein-coupled receptors, with six new crystallographic structures emerging in the last 3½ years. Slowly, it is becoming an accepted method with a proven record and convincing generality. However, it is not a method that is used in every membrane structural biology laboratory and that is unfortunate. The reluctance in adopting it is attributable, in part, to the anticipated difficulties associated with handling the sticky viscous cubic mesophase in which crystals grow. Harvesting and collecting diffraction data with the mesophase-grown crystals is also viewed with some trepidation. It is acknowledged that there are challenges associated with the method. However, over the years, we have worked to make the method user-friendly. To this end, tools for handling the mesophase in the pico- to nano-litre volume range have been developed for efficient crystallization screening in manual and robotic modes. Glass crystallization

[1]*To whom correspondence should be addressed (email martin.caffrey@tcd.ie).*

plates have been built that provide unparalleled optical quality and sensitivity to nascent crystals. Lipid and precipitant screens have been implemented for a more rational approach to crystallogenesis, such that the method can now be applied to a wide variety of membrane protein types and sizes. In the present article, these assorted advances are outlined, along with a summary of the membrane proteins that have yielded to the method. The challenges that must be overcome to develop the method further are described.

Introduction

The 13 January 2011 issue of the journal *Nature* includes three papers reporting on the structure and function of medically and pharmaceutically important GPCRs (G-protein-coupled receptors) [1–3]. Structures were determined using MX (macromolecular crystallography) and in two of these papers, the lipidic cubic phase method was used to generate diffraction quality crystals. The latter, so-called *in meso* crystallogenesis method, is the focus of the present article.

The structures described in the *Nature* papers refer to how the atoms that constitute these receptor molecules are arranged in three-dimensional space. This structural information provides invaluable insight into the detailed workings and interactions of these macromolecules and how they might be regulated for the betterment of humankind. It calls to mind the seminal X-ray work of J.D. Bernal and W.H. Bragg carried out at the University of Cambridge almost a century ago on the structure of the carbon allotropes graphite and diamond, where the relationship between structure and material properties was brought into stark relief. Bernal extended the structure–function principle into the realm of biomacromolecules, contributing in no small way to the development of what we recognize today as Structural Biology [4].

The Membrane Structural and Functional Biology (MS&FB) group at Trinity College Dublin follows in the Bernal tradition of seeking to decipher and to understand the intricacies of the biomolecular structure–function relationship. Here, the focus is on the biological membrane, a bimolecular lipid leaflet in and on which are situated a multitude of proteins and other molecules. The approach taken in the group is a holistic one in that care is taken to perform measurements and to evaluate results in the context of the native biomembrane. Therefore, for this reason, particular attention is devoted to the role played by lipids in the structure, activity and interactions of membrane proteins, a theme emphasized at this Biochemical Society Annual Symposium.

I began my career working with membrane proteins involved in lipid metabolism and calcium transport [5,6]. From early on, I recognized the critical role that lipid-phase properties were likely to play in membrane protein function. The dearth of knowledge in the area at the time induced me to embark on an extended period of investigation devoted almost exclusively to establishing the relationship between lipid molecular structure and mesophase or liquid crystalline behaviour and microstructure [7]. One of the goals of that work was to establish rules for rationally designing lipids of defined functionality. This objective was realized after many years and on the back of a lipid synthesis and mesophase characterization programme established in the MS&FB group. It has provided the community with a suite of lipids having desirable properties for membrane protein crystallization and other applications (see below).

With a solid training in membrane protein biochemistry and biophysics and a strong background in lipid-phase science my group and I were nicely poised to enter the field of macromolecular crystallography as applied to membrane proteins. The opportunity to do just that, and to repot scientifically, came when E. Landau and J. Rosenbusch at the Biocentre in Basel reported that the light-driven transmembrane proton pump bacteriorhodopsin crystallized in a cubic mesophase when treated with phosphate at molar concentrations [8]. At that time, the group had built up considerable experience working with lipidic mesophases, the cubic phase in particular, and had mapped out the temperature–composition phase diagram for the mono-olein–water system (Figure 1) which was directly relevant to the bacteriorhodopsin crystallogenesis work [8]. Collaboration between the group in Basel and the MS&FB group was established which led to exchanges of materials, methods, equipment, knowledge and personnel, and to work on the mechanism of membrane protein crystallization *in meso* [9,10].

The *in meso* method was particularly appealing to my aforementioned holistic sensibilities because it offered the prospect that crystallization was taking place from within a lipid bilayer, akin to the native environment encountered in a biomembrane. This was in contrast with the more traditional '*in surfo*' methods for crystallizing membrane proteins [7,10a] available at the time which involved using potentially destabilizing surfactant micelles. Altogether, the method and the opportunities it presented made it an attractive direction in which to channel our research efforts which have three major themes. First, to decipher the molecular basis of *in meso* crystallogenesis. Secondly, to automate and miniaturize the method and to make it more user-friendly and generally accessible. And, thirdly, to use the method to solve the structures of membrane proteins that are critical to human health. In what follows, the origins of the *in meso* method, its development as a high-throughput technique, the membrane proteins that have yielded to it, and the challenges ahead are described.

The *in meso* method: how it works and the challenges it presents

Setting up an *in meso* crystallization trial is straightforward. Typically, it involves combining two parts protein solution with three parts lipid at 20°C [11,12]. The lipid most commonly used is the MAG (monoacylglycerol) mono-olein. According to the mono-olein–water phase diagram (Figure 1) [13], and assuming there is no major influence on phase behaviour of the protein solution components, this mixing process should generate, by spontaneous self-assembly,

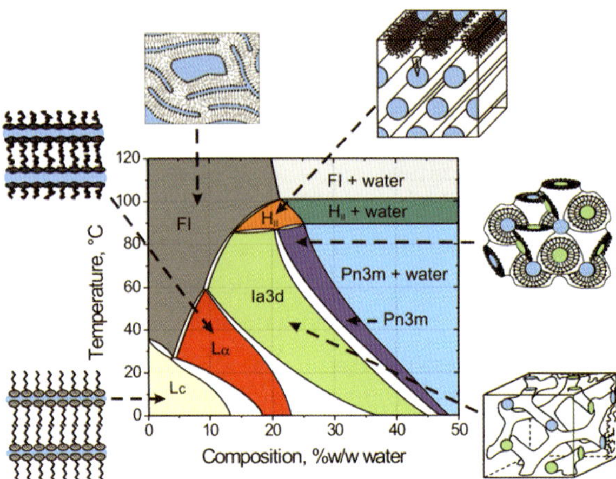

Figure 1. Temperature–composition phase diagram of the mono-olein–water system determined under 'conditions of use' in the heating and cooling directions from 20°C
A cartoon representation of the various phase states is included in which coloured zones represent water. The liquid crystalline phases below ~17°C are metastable [13]. Abbreviations: Fl, fluid isotropic phase; H_{II}, inverted hexagonal phase; $L\alpha$, lamellar liquid crystalline phase; Lc, lamellar crystal phase. Figure reproduced from [24] with permission.

the cubic mesophase (Figure 2) at, or close to, full hydration. The original method for mixing lipid and protein solution involved multiple cumbersome centrifugations in small glass tubes. Harvesting crystals required cutting the tubes and searching for small crystals through curved glass, which was not easy, and required experience, time and patience.

The cubic phase is sticky and viscous in the manner of thick toothpaste. As such, it is not easy to handle. In the course of our earlier lipid-phase science work, we had developed tools and procedures for manipulating such refractory materials. One of these, the syringe mixing device [14], was ideally suited to the task of combining microlitre volumes of mono-olein with membrane protein solution in a way that produces protein-laden mesophase for direct use in crystallization trials with minimal waste. The mixer consists of two Hamilton microsyringes connected by a narrow bore coupler. Lipid is placed in one syringe, and protein solution in the other. Mixing is achieved by repeatedly moving the contents of the two syringes back and forth through the coupler [11]. The coupler is replaced by a needle for convenient dispensing of the homogeneous mesophase into wells of custom-designed glass sandwich crystallization plates [15,16]. Precipitant solutions of various compositions are placed over the mesophase and the wells are sealed with a coverglass. The plates are incubated at 20°C and monitored for crystal growth. Optical quality is the best it can be given that the mesophase is held between two glass plates and the mesophase itself is transparent. This means that crystals, just a few micrometres in size, can be seen readily using microscopy, whether or not the proteins are coloured. The use of cross-polarizers enhances the visibility of small

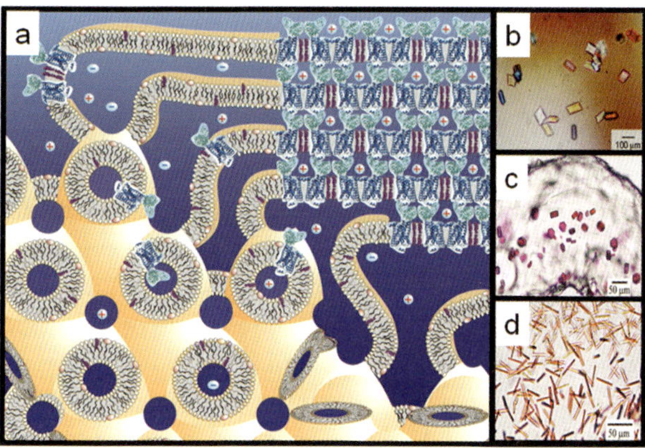

Figure 2. *In meso* crystallization model and crystals
(**a**) Schematic representation of the events proposed to take place during the crystallization of an integral membrane protein from the lipidic cubic mesophase. The process begins with the protein reconstituted into the highly curved bilayers of the bicontinuous cubic phase (bottom left quadrant). Added precipitants shift the equilibrium away from stability in the cubic membrane. This leads to phase separation, wherein protein molecules diffuse from the continuous bilayered reservoir of the cubic phase by way of a sheet-like or lamellar portal (upper left quadrant) to lock into the lattice of the advancing crystal face (upper right quadrant). Salt (positive and negative signs) facilitates crystallization, in part, by charge screening. Co-crystallization of the protein with native or added lipid (cholesterol) is shown in this illustration. As much as possible, the dimensions of the lipid (light yellow oval with tail), detergent (pink oval with tail), native membrane or added lipid (purple), protein (blue; β_2-adrenergic receptor–T4 lysozyme; PDB code 2RH1), and bilayer and aqueous channels (dark blue) have been drawn to scale. The lipid bilayer is approximately 40 Å thick. Crystals of (**b**) BtuB, (**c**) bacteriorhodopsin and (**d**) light-harvesting complex II growing *in meso*. Reprinted, with permission, from the *Annual Review of Biophysics* [10a], Volume 38, © 2009 by Annual Reviews.

crystals which usually appear birefringent in a dark background; the cubic phase itself is optically isotropic and non-birefringent. An added feature of the glass sandwich plates is that the double-sided tape used to create the wells provides almost hermetic sealing, ensuring minimal changes in well composition during the course of trials that can last for months. Step-by-step instructions, complete with an online video demonstration of the entire *in meso* crystallization process just described, have been published [11,12].

The *in meso* robot

The protocol just described refers to the manual mode of setting up crystallization trials. Accurate and precise delivery of the protein-laden mesophase in volumes that range from pico- to micro-litres was made possible by use of an inexpensive repeat dispenser in combination with differently sized microsyringes [17,18]. The smaller volumes means that the *in meso* method works with minuscule quantities of target protein. Thus extensive crystallization trials can be set up with just a few micrograms of valuable membrane protein, making the *in meso* method one of the most efficient in terms of protein requirement.

Whereas the repeat dispenser greatly facilitated the *in meso* method, it was still a manual set-up, with limits to the numbers of trials that any one person could comfortably set up at a sitting. The need to automate the process was obvious. With the assistance of A. Peddi and Y. Zheng, engineers at The Ohio State University where the original work was carried out, we were able to perform a proof-of-principle robotics exercise employing LabView-controlled motorized translation stages operating and supporting a microsyringe and a crystallization plate. With it, we demonstrated that the viscous mesophase could be dispensed automatically and wells were filled in such a way that eventually yielded crystals [19]. This was enough to secure funding for a robot which was custom-designed and built to our specifications.

The *in meso* robot has two arms programmed to move simultaneously. One dispenses the viscous protein-laden mesophase, while the other dispenses precipitant. Typical volumes used are 50 nl of mesophase (consisting of 20 nl of protein solution and 30 nl of mono-olein) and 800 nl of precipitant solution. Custom 96-well glass sandwich plates were designed which take approximately 6 min to fill using an eight-tip robot. The robot enables the precise and accurate setting up of *in meso* crystallization trials in high-throughput mode and, if required, under challenging conditions of reduced temperature and controlled lighting. Given the *in meso* robot's success, several are currently in use in laboratories throughout the world. A variant on the original design, where tip alignment is done automatically and where precipitant is handled by disposable tips, is now commercially available. It would appear to represent an important advance in simplifying the *in meso* set up and making the method user-friendly.

With the success that the *in meso* method has had, it perhaps is not unexpected to find products appearing on the market in support of this novel crystallogenesis approach. In addition to the *in meso* robots, these include a number of precipitant screen kits, glass and plastic sandwich plates, and a plate that comes complete with lipid-coated wells. The vendors indicate that the latter can be used with a liquid-dispensing robot for protein solution delivery first and precipitant post-swelling.

Mesophase compatibility with protein solution components

As alluded to above, what happens during *in meso* crystallization is intimately tied up with mesophase behaviour [20]. The working hypothesis for how nucleation comes about begins with the protein reconstituting into the continuous bilayer of the cubic phase (Figure 2). Precipitant is added which triggers local formation of a lamellar phase into which the protein preferentially partitions and concentrates in a process that leads to nucleation and crystal growth. Experimental evidence in support of aspects of this model has been reported [20,21].

Experience built up over several years working with the *in meso* method suggests that the mesophase behaviour observed during the course of crystallization mimics that of the mono-olein–water system (Figure 1). The implication therefore is that the protein solution has minimal effect on the phase behaviour of the hosting lipidic mesophase into which the protein is reconstituted. That solution, along with the target protein, typically includes lipid, detergent, buffers and salt at a minimum. Other components such as glycerol, thiol-group-specific reagents or denaturants are not uncommon. Each of these can have an impact on phase behaviour and, by extension, on the outcome of a crystallization trial. In the interests of learning about component compatibility, the sensitivity of the mono-olein–water cubic phase system to their inclusion has been evaluated. Our findings indicate that the default cubic mesophase is remarkably resilient and retains its phase identity in the presence of a vast array of different additives. These include glycerolipids, cholesterol, free fatty acids, detergents, denaturants, glycerol and thiol-group-specific reagents [22–30]. Of course, for each there is a concentration beyond which the cubic phase is no longer stable nor, indeed, useful for crystallogenesis. In most cases, these limits have been identified.

Occasionally, the concentration of a protein solution component is not known exactly. Detergent is a case in point. This poses a problem because, if there is too much detergent, the bulk lamellar phase may form, but it will not support crystallization [22,31]. It may also be that a new detergent is being used whose compatibility with the cubic phase is not known. In this case, a small amount of the buffer used to solubilize the protein or the protein solution itself can be used to prepare mesophase. The physical texture, appearance between crossed polarizers, or SAXS (small-angle X-ray scattering) behaviour of the mesophase will indicate which phase has been accessed. If, for example, it is a lamellar phase that forms, suggesting too much detergent, then another purification step where its concentration in the final protein solution is reduced may be enough to solve the problem. We have encountered situations with bacteriorhodopsin where the particular preparation ended up having an excess of detergent. The mesophase first formed was lamellar, but, when it was used in combination with certain precipitants, a transition back to the cubic phase was induced which went on to support crystal growth [20].

Screen solution compatibility

As noted, *in meso* crystallization relies upon a bicontinuous mesophase which acts as a reservoir to feed protein into nucleation sites and for crystal growth (Figure 2). The crystallization screening process requires that chemical space be interrogated over wide limits to find conditions that support crystal growth. In the screening process the protein-laden mesophase is therefore typically exposed to precipitant solutions that encompass hundreds, perhaps even thousands, of different chemical compositions. Screen solution components typically include buffers that cover a wide pH range, polymers, salts, small organics, detergents, apolar solvents, amphiphiles, etc., and all at different concentrations. Each component can potentially destabilize the mesophase. In a separate study using SAXS, we examined the compatibility of the default mono-olein–water cubic phase with various commonly used precipitant screen solutions [25]. What we found was hardly surprising. Compatibility was temperature-dependent and the usual suspects, which included organic solvents, destroyed the cubic phase, rendering these screen solutions effectively useless. A goal of the study was to design screens that were mesophase-friendly. However, that goal was never pursued; instead, we have opted for the convenience of commercial screen kits mindful of the fact that certain conditions are not relevant. As a result, certain kits are simply not used because they contain too few conditions that are compatible with the cubic phase.

Sponge phase

During the course of mesophase-compatibility studies, we noticed that certain screen components caused the cubic phase to 'swell' and, under certain conditions, to form what is referred to as the sponge phase. The latter evolves from the cubic phase as a result of the 'spongifying' component lowering bilayer interfacial curvature, thereby enabling the mesophase to imbibe more lyotrope (aqueous solution). This is revealed in the SAXS pattern where the lattice parameter of the cubic phase rises. Eventually, the mesophase loses order and the low-angle diffraction pattern becomes diffuse. Fortunately, the sponge phase retains its bicontinuity and, as a result, can support *in meso* crystallogenesis [20,24,32]. One advantage of the sponge phase is that its aqueous channels are dilated. Thus proteins with large extramembrane domains should be accommodated in and amenable to crystallogenesis from the sponge phase. Furthermore, the reduced interfacial curvature is likely to facilitate more rapid and long-range diffusion within the lipid bilayer. Since net movement of protein from the bulk mesophase reservoir to the nucleation and growth sites is a requirement for crystallization, this effect alone should contribute to improved crystallization. Interestingly, many of the proteins that have yielded to the *in meso* method have been crystallized under conditions that favour sponge phase formation ([33] and see http://www.mpdb.tcd.ie).

Reflecting the utility of the sponge phase for *in meso* crystallogenesis, a number of commercial screening kits now include spongifiers such as poly(ethylene glycol), jeffamine and butanediol. Some of these provide a preformed sponge phase to which the protein solution is added directly. We continue to use the original method that involves an active protein reconstitution step and where the entire crystallization screen space is available for sampling.

Rational lipid design for low-temperature crystallogenesis

As described in the Introduction, the MS&FB group has devoted considerable time and effort to establishing the rules for rationally designing lipids with specific end uses. One such application concerned the development of a host lipid for use in *in meso* crystallogenesis at low temperatures. Certain proteins are labile and require handling at low temperatures. The problem with the *in meso* method, in the default mode at least, is that it relies upon mono-olein as the hosting lipid. The cubic phase formed by mono-olein is not stable below ~17°C [13] and performing crystallization trials in a cold room at 4–6°C is risky. For this low-temperature application a *cis*-mono-unsaturated MAG, 7.9 MAG, was therefore designed, using the rules referred to above. The target MAG was synthesized and purified in-house and its phase behaviour mapped out using SAXS [34]. As designed, it produced the cubic phase stable in the range 6–85°C. 7.9 MAG has been used in the crystallization of a number of membrane proteins by the MS&FB group. The objective now is to make it, along with other synthetic MAGs (see below), available to the community by way of a commercial vendor.

The word 'risky' was used in the previous paragraph when referring to low-temperature crystallization with mono-olein as the hosting lipid. This reflects the fact that it is possible to do successful *in meso* work with mono-olein at 4°C provided the system undercools. Fortunately, the cubic phase is noted for this capacity [13,35] and we regularly perform successful crystallization trials with mono-olein in the 4–17°C range. As expected, occasionally under these metastable conditions, the mesophase will convert into the lamellar crystalline or solid phase which is useless as far as crystallization is concerned.

Sugar-phytane lipids have been synthesized that form the fully hydrated cubic phase in the 10–70°C range [36] and that might find application for *in meso* crystallization at reduced temperatures.

Additive lipid screening

Early on in the development of the *in meso* method, I recognized that mono-olein, as the lipid used to create the hosting mesophase, is a most uncommon membrane lipid. The sense was that this MAG might rightly be regarded as foreign by certain target proteins and cause them to destabilize. One possible solution was to use a natural membrane lipid that would form the requisite cubic phase at 20°C. None was available. An alternative was to use mono-olein as the hosting lipid and to augment it with typical membrane lipids, thereby creating a more native-like environment. Accordingly, the carrying capacity of the mono-olein cubic phase for a number of different lipids was established using SAXS [23]. This amounted to approximately 20 mol% in the case of phosphatidylcholine, phosphadidylethanola-mine and cholesterol, with lesser amounts of phosphatidylserine and cardiolipin being accommodated. The approach of using additive lipids has had spectacular successes in the GPCR field where cholesterol doping of the cubic phase was critical to the production of structure-grade crystals [1,2,37–41].

Host lipid screening

Mono-olein was the first lipid used for *in meso* crystallogenesis. From the outset, it was recognized that this one lipid may not work with all target membrane proteins. These, in turn, come from a variety of native membranes which differ in lipid composition, surface charge and packing density, fluidity and polarity profile, bilayer thickness, intrinsic curvature, etc. Thus having a range of MAGs that differed in acyl chain characteristics with which to screen was deemed important. Using principles of rational design, a number of suitable MAGs were identified with the requirement that they form the cubic phase at 20°C under conditions of full hydration. A number of lipids meeting this specification have been synthesized and characterized in-house. They constitute a successful hosting lipid screen by the MS&FB group. With several targets, that include β-barrels, α-helical proteins (see below) and an integral peptide antibiotic, crystals have been grown by the *in meso* method using these alternative MAGs [34,42–44]. In a number of cases, mono-olein either failed to produce crystals or the crystals it did produce were not of diffraction quality. It was only when MAGs from the hosting lipid screen were used that structure-grade crystals were obtained. The intent now is to make these novel MAGs available to the community through a suitable supplier.

In meso structures

At the time of writing, the *in meso* method accounts for almost 70 records in the PDB that relate to integral membrane proteins and peptides ([33] and see http://www.mpdb.tcd.ie). This corresponds to ~10% of all published membrane protein structures representing six distinct membrane protein types (Figure 3). With successes that include bacterial rhodopsins, light-harvesting complex II, photosynthetic reaction centres, β-barrels, GPCRs and an integral membrane peptide, the method has a record of versatility and range. In the MS&FB group, a multi-subunit cytochrome oxidase structure is in the final stages of refinement, which further attests to the generality of the method (J. Lyons, D. Aragão, T. Soulimane and M. Caffrey, unpublished work; PDB code 2YEV). Each of these membrane protein types represents bigger families, the members of which become suitable candidates for *in meso* crystallogenesis. The GPCR family is a case in point, with almost 800 distinct GPCRs coded for in the human genome alone. The *in meso* method, in combination with the necessary protein engineering and receptor stabilization strategies, is therefore now poised to contribute to the generation of GPCR structures in, what amounts to, production-line fashion. Evidence in support of this statement is the recent

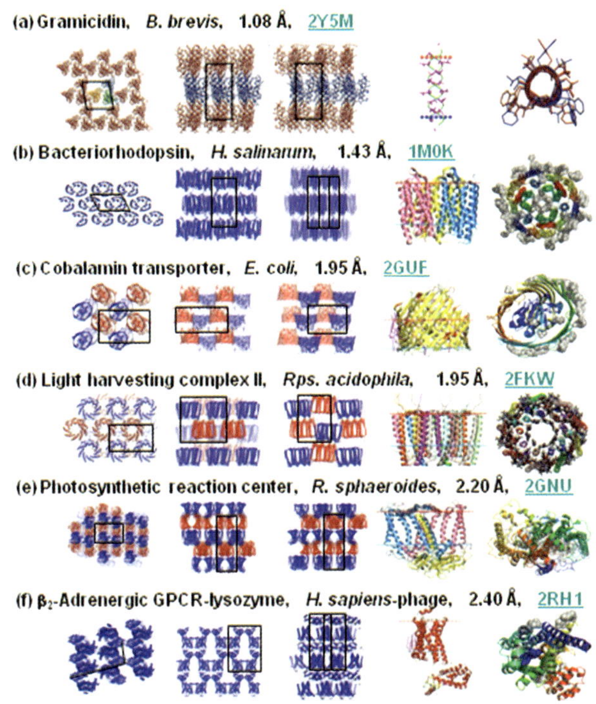

Figure 3. *In meso* crystal packing arrangement and molecular structures of membrane proteins. The packing arrangement is shown in columns 1–3 within each panel. Expanded views of individual proteins or oligomers are shown in columns 4 and 5. The view in columns 1 and 5 represents a projection along the stacking axis. The views in columns 2–4 are from within the plane of the stacked lamellae along the two other unit cell axes. In columns 1–3, black outlines are projections of the unit cell. Images in column 4 are from the Orientations of Proteins in Membranes (OPM) database (http://opm.phar.umich.edu/), in which the red and blue horizontal lines define the hydrophobic thickness of the protein. Proteins are identified by name, source organism, resolution and PDB code. Currently, *in meso* records in the MPDB (http://www.mpdb.tcd.ie) number 44 for bacterial rhodopsins, six for the photosynthetic reaction centres, 13 for GPCR–T4 lysozyme chimaeras, three for gramicidin, two for β-barrels and one for light-harvesting complex II. The examples shown represent the highest resolution available for each of the six protein types [33]. *B. brevis*, *Bacillus brevis*; *H. salinarum*, *Halobacterium salinarum*; *E. coli*, *Escherichia coli*; *Rps. acidophila*, *Rhodopseudomonas acidophila*; *R. sphaeroides*, *Rhodobacter sphaeroides*; *H. sapiens*, *Homo sapiens*.

spate of receptor structures courtesy of the *in meso* method. We can only hope for the same degree of success with other membrane protein families.

The further development of the *in meso* crystallogenesis approach is an important goal for members of the MS&FB group. One direction this has taken recently concerns the utility of the method with small membrane proteins. A separate analysis performed using a model cubic phase under restricted conditions indicated that suitable targets would need to include at least five transmembrane helices [45]. Our experience with the sponge-phase variant of the cubic phase suggested otherwise. Accordingly, the utility of the method with a 'mini-protein', the pentadecapeptide antibiotic linear gramicidin was investigated. It worked remarkably well, providing a structure with a resolution better than 1.1 Å (1 Å = 0.1 nm) [44,46]. This result is significant because it highlights the utility of the method with proteins having small transmembrane domains which abound in Nature.

The Membrane Protein Data Bank, statistics online

Further details regarding the structure and function of integral, anchored and peripheral membrane proteins are available online in a convenient and searchable database, the MPDB (Membrane Protein Data Bank) at http://www.mpdb.tcd.ie [33]. Although records in the MPDB are obtained from the PDB, the former only includes entries for membrane proteins. Statistical analyses on the contents of the database can be performed and viewed online. Examples include detergents used for membrane protein structure work and the number of structures published annually by method.

Prospects

The *in meso* method burst on the scene a decade and a half ago. It was received with great anticipation for what it would deliver; perhaps it was to be the panacea. However, output in the early years was limited to naturally abundant bacterial α-helical proteins bedecked with stabilizing and highly coloured prosthetic groups ([33] and see http://www.mpdb.tcd.ie). The perceived restricted range, coupled to the challenges associated with handling the sticky and viscous cubic mesophase, meant that subsequent interest in the method waned. This was countered to some degree with the introduction of the *in meso* robot, a growing understanding for how the method worked at a molecular level, and a continued demonstration of the method's general applicability. However, interest in the method has rocketed of late, with the success that it has had in the GPCR field [1,2,37–41].

Improvements are, of course, needed if the method is to have longevity. Critically, the specialized materials and supplies upon which the method relies must be made more generally available and the method itself must be made user-friendly and routine. New and improved *in meso* robots available on the market are tackling the user-friendliness issue. Workshops that involve hands-on demonstrations contribute to making the method more accessible. I have been active in this area for several years now, with the latest workshop held as a part of the ICCBM13 meeting in Dublin in September 2010 (http://www.iccbm13.ie) [47]. There, 36 students were trained in the practicalities and finer aspects of *in meso* crystallogenesis. A video demonstration of the method is available online [11].

Developments are needed in the area of crystal identification. Optical clarity is of the highest quality with the glass sandwich plates currently in use and this provides for ready detection of colourless micrometre-sized crystals in normal light and between crossed polarizers. Detection by UV fluorescence is particularly powerful and convenient for tryptophan-containing proteins. Fluorescence labelling [48] is also a route worth considering for the sensitive detection of early hits. SONICC (second-order non-linear optical imaging of chiral crystals) is a novel approach introduced by G. Simpson. It has been shown to sensitively and selectively detect membrane protein crystals growing *in meso* [49].

Recovering crystals from the mesophase for data collection is a non-trivial undertaking [12]. This is especially true when harvesting is done directly from glass sandwich plates. Typically, a glass cutter is used to open the well and to expose the mesophase. Teasing out and harvesting the crystal for immediate cryocooling is most conveniently done with a cryoloop. This is a slow, painstaking and cumbersome process especially if it must be done in a cold room and/or under subdued light. This whole area of harvesting calls out for innovation.

Data collection at the synchrotron is not exactly straightforward either. Given that *in meso*-grown crystals tend to be small, a mini-synchrotron X-ray beam is required. Oftentimes, the crystal of interest is hidden from view in a bolus of mesophase on the cryoloop. This means that locating the crystal and centring it requires rounds of diffraction rastering with a beam of progressively smaller size [50]. This same approach is used to advantage in finding the best diffracting part of a crystal. Locating crystals and centring based on X-ray fluorescence from heavy atoms in the sample is in development (D. Aragão, M. Becker, D. Li, M. Hilgart, J. Lyons, D. Yoder, S. Stepanov, R. Fischetti and M. Caffrey, unpublished work). Effective and efficient rastering is recognized now as an important feature of the latest MX beamlines at synchrotron facilities worldwide and steady improvements in the rastering process are being made. *In situ* screening and data collection are other areas under investigation.

The structures solved to date using *in meso*-grown crystals have relied on molecular replacement for phasing. However, new structures are in the offing that require experimental phasing. In our hands at least, this is proving to be a challenge. Several targets have been tackled using selenomethionine labelling and co-crystallization and soaking with heavy atoms without success. The problem derives from a low anomalous signal-to-noise ratio due to a combination of background scatter from adhering mesophase and the need to work with small and sometimes poorly diffracting radiation-sensitive crystals. This part of the *in meso* pipeline is in need of work.

Finally, the method should begin to be used with really big proteins and complexes. The sponge phase [24], with its open aqueous channels and flatter bilayer, should prove particularly useful in this regard. Using it in combination with novel hosting and additive lipid screens [23,42] will go a long way towards producing crystals and ultimately high-resolution structures where interactions that are integral to human health are revealed.

Acknowledgements

There are many who contributed to this work and most are from my own group, both past and present members. To all I extend my warmest thanks and appreciation.

Funding

This work was supported in part by Science Foundation Ireland [grant number 07/IN.1/B1836], Framework Programme 7 COST Action CM0902 and the National Institutes of Health [grant numbers GM75915, P50GM073210 and U54GM094599].

References

1. Rasmussen, S.G.F., Choi, H.-J., Fung, J.J., Pardon, E., Casarosa, P., Chae, P.S., DeVree, B.T., Rosenbaum, D.M., Thian, F.S., Kobilka, T.S. et al. (2011) Structure of a nanobody-stabilized active state of the β_2 adrenoceptor. *Nature* **469**, 175–180
2. Rosenbaum, D.M., Zhang, C., Lyons, J.A., Holl, R., Aragao, D., Arlow, D.H., Rasmussen, S.G.F.R., Choi, H.-J., DeVree, B.T., Sunahara, R.K. et al. (2011) Structure and function of an irreversible agonist–β_2 andrenoreceptor complex. *Nature* **469**, 236–240
3. Warne, A., Moukhametzianov, R., Baker, J.G., Nehme, R., Edwards, P.C., Leslie, A.G.W., Schertler, G.F.X. & Tate, C.G. (2011) The structural basis for agonist and partial agonist action on a β_1-adrenergic receptor. *Nature* **469**, 241–244
4. Caffrey, M. (2007) J.D. Bernal and the genesis of structural biology. *J. Physics Conf. Ser.* **57**, 17–28
5. Caffrey, M. & Kinsella, J.E. (1977) Growth and acyltransferase activity of rabbit mammary gland during pregnancy and lactation. *J. Lipid Res.* **18**, 44–52
6. Caffrey, M. & Feigenson, G.W. (1981) Fluorescence quenching in model membranes: the relationship between Ca^{2+}-ATPase enzyme activity and the affinity of the protein for phosphatidylcholines with different acyl chain characteristics. *Biochemistry* **20**, 1949–1961
7. Caffrey, M. (2003) Membrane protein crystallization. *J. Struct. Biol.* **142**, 108–132
8. Landau, E.M. & Rosenbusch, J.P. (1996) Lipidic cubic phases: a novel concept for the crystallization of membrane proteins. *Proc. Natl. Acad. Sci. U.S.A.* **93**, 14532–14535
9. Caffrey, M. (2000) A lipid's eye view of membrane protein crystallization in mesophases. *Curr. Opin. Struct. Biol.* **10**, 486–497
10. Nollert, P., Qiu, H., Caffrey, M., Rosenbusch, J.P. & Landau, E.M. (2001) Molecular mechanism for the crystallization of bacteriorhodopsin in lipidic cubic phases. *FEBS Lett.* **504**, 179–186

10a. Caffrey, M. (2009) Crystallizing membrane proteins for structure determination: use of lipidic mesophases. *Annu. Rev. Biophys.* **38**, 29–51

11. Caffrey, M. & Porter, C. (2010) Crystallizing membrane proteins for structure determination using lipidic mesophases. J. Vis. Exp., http://www.jove.com/index/details.stp?id=1712, doi:10.3791/1712
12. Caffrey, M. & Cherezov, V. (2009) Crystallizing membrane proteins in lipidic mesophases. *Nat. Protoc.* **4**, 706–731
13. Qiu, H. & Caffrey, M. (2000) Phase diagram of the monoolein/water system: metastability and equilibrium aspects. *Biomaterials* **21**, 223–234
14. Cheng, A., Hummel, B., Qiu, H. & Caffrey, M. (1998) A simple mechanical mixer for small viscous lipid-containing samples. *Chem. Phys. Lipids* **95**, 11–21

15. Cherezov, V. & Caffrey, M. (2003) Nano-volume plates with excellent optical properties for fast, inexpensive crystallization screening of membrane proteins. *J. Appl. Crystallogr.* **36**, 1372–1377
16. Cherezov, V., Peddi, A., Muthusubramaniam, L., Zheng, Y.F. & Caffrey, M. (2004) A robotic system for crystallizing membrane and soluble proteins in lipidic mesophasesActa Crystallogr. *Sect. D Biol. Crystallogr.* **60**, 1795–1807
17. Cherezov, V. & Caffrey, M. (2005) A simple and inexpensive nanoliter-volume dispenser for highly viscous materials used in membrane protein crystallization. *J. Appl. Crystallogr.* **38**, 398–400
18. Cherezov, V. & Caffrey, M. (2006) Picoliter-scale crystallization of membrane proteins. *J. Appl. Crystallogr.* **39**, 604–609
19. Cherezov, V., Peddi, A., Muthusubramaniam, L., Zheng, Y.F. & Caffrey, M. (2004) A robotic system for crystallizing membrane and soluble proteins in lipidic mesophasesActa Crystallogr. *Sect. D Biol. Crystallogr.* **60**, 1795–1807
20. Caffrey, M. (2008) On the mechanism of membrane protein crystallization in lipidic mesophases. *Cryst. Growth Des.* **8**, 4244–4254
21. Cherezov, V. & Caffrey, M. (2007) Membrane protein crystallization in lipidic mesophases: a mechanism study using X-ray microdiffraction. *Faraday Discuss.* **136**, 188–205
22. Ai, X. & Caffrey, M. (2000) Membrane protein crystallization in lipidic mesophases: detergent effects. *Biophys. J.* **79**, 394–405
23. Cherezov, V., Clogston, J., Misquitta, Y., Abdel-Gawad, W. & Caffrey, M. (2002) Membrane protein crystallization *in meso*: lipid type-tailoring of the cubic phase. *Biophys. J.* **83**, 3393–3407
24. Cherezov, V., Clogston, J., Papiz, M. & Caffrey, M. (2006) Room to move: crystallizing membrane proteins in swollen lipidic mesophases. *J. Mol. Biol.* **357**, 1605–1618
25. Cherezov, V., Fersi, H. & Caffrey, M. (2001) Crystallization screens: compatibility with the lipidic cubic phase for *in meso* crystallization of membrane proteins. *Biophys. J.* **81**, 225–242
26. Clogston, J. & Caffrey, M. (2005) Controlling release from the cubic phase: amino acids, peptides, proteins and nucleic acids. *J. Controlled Release* **107**, 97–111
27. Clogston, J., Graciun, G., Hart, D.J. & Caffrey, M. (2005) Controlling release from the lipidic cubic phase by selective alkylation. *J. Controlled Release* **102**, 441–461
28. Liu, W. & Caffrey, M. (2005) Gramicidin structure and disposition in highly curved membranes. *J. Struct. Biol.* **150**, 23–40
29. Liu, W. & Caffrey, M. (2006) Interactions of tryptophan, tryptophan peptides and tryptophan alkyl esters at curved membrane interfaces. *Biochemistry* **45**, 11713–11726
30. Cherezov, V., Yamashita, E., Liu, W., Zhalnina, M., Cramer, W.A. & Caffrey, M. (2006) *In meso* structure of the cobalamin transporter, BtuB, at 1.95 Å resolution. *J. Mol. Biol.* **364**, 716–734
31. Misquitta, Y. & Caffrey, M. (2003) Detergents destabilize the cubic phase of monoolein: implications for membrane protein crystallization. *Biophys. J.* **85**, 3084–3096
32. Wöhri, A.B., Johansson, L.C., Wadsten-Hindrichsen, P., Wahlgren, W.Y., Fischer, G., Horsefield, R., Katona, G., Nyblom, M., Oberg, F., Young, G. et al. (2008) A lipidic-sponge phase screen for membrane protein crystallization. *Structure* **16**, 1003–1009
33. Raman, P., Cherezov, V. & Caffrey, M. (2006) The Membrane Protein Data Bank. *Cell. Mol. Life Sci.* **63**, 36–51
34. Misquitta, Y., Cherezov, V., Havas, F., Patterson, S., Mohan, J.M., Wells, A.J., Hart, D.J. & Caffrey, M. (2004) Rational design of lipid for membrane protein crystallization. *J. Struct. Biol.* **148**, 169–75
35. Briggs, J. & Caffrey, M. (1994) The temperature-composition phase diagram of mono-myristolein in water: equilibrium and metastability aspects. *Biophys. J.* **66**, 573–587
36. Hato, M., Minamikawa, H., Salkar, R.A. & Matsutani, S. (2004) Phase behavior of phytanyl-chained akylglycoside/water systems. *Prog. Colloid Polym. Sci.* **123**, 56–60
37. Rosenbaum, D.M., Cherezov, V., Hanson, M.A., Rasmussen, S.G., Thian, F.S., Kobilka, T.S., Choi, H.J., Yao, X.J., Weis, W.I., Stevens, R.C. & Kobilka, B.K. (2007) GPCR engineering yields high-resolution structural insights into β_2-adrenergic receptor function. *Science* **318**, 1266–1273

38. Cherezov, V., Rosenbaum, D.M., Hanson, M.A., Rasmussen, S.G., Thian, F.S., Kobilka, T.S., Choi, H.J., Kuhn, P., Weis, W.I., Kobilka, B.K. & Stevens, R.C. (2007) High-resolution crystal structure of an engineered human β_2-adrenergic G protein-coupled receptor. *Science* **318**, 1258–1265
39. Jaakola, V.P., Griffith, M.T., Hanson, M.A., Cherezov, V., Chien, E.Y., Lane, J.R., Ijzerman, A.P. & Stevens, R.C. (2008) The 2.6 angstrom crystal structure of a human A2A adenosine receptor bound to an antagonist. *Science* **322**, 1211–1217
40. Wu, B., Chien, E.Y., Mol, C.D., Fenalti, G., Liu, W., Katritch, V., Abagyan, R., Brooun, A., Wells, P., Bi, F.C. et al. (2010) Structures of the CXCR4 chemokine GPCR with small-molecule and cyclic peptide antagonists. *Science* **330**, 1066–1071
41. Chien, E.Y., Liu, W., Zhao, Q., Katritch, V., Han, G.W., Hanson, M.A., Shi, L., Newman, A.H., Javitch, J.A., Cherezov, V. & Stevens, R.C. (2010) Structure of the human dopamine D3 receptor in complex with a D2/D3 selective antagonist. *Science* **330**, 1091–1095
42. Li, D., Lee, J. & Caffrey, M. (2011) Crystallizing membrane proteins in lipidic mesophases: a host lipid screen. *Cryst. Growth Des.* **11**, 530–537
43. Misquitta, L.V., Misquitta, Y., Cherezov, V., Slattery, O., Mohan, J.M., Hart, D., Zhalnina, M., Cramer, W.A. & Caffrey, M. (2004) Membrane protein crystallization in lipidic mesophases with tailored bilayers. *Structure* **12**, 2113–2124
44. Hoefer, N., Aragao, D., Lyons, J.M. & Caffrey, M. (2011) Membrane protein crystallization in lipidic mesophases: hosting lipid effects on the crystallization and structure of a transmembrane peptide. *Cryst. Growth Des.*, doi:10.1021/cg101384p
45. Grabe, M., Neu, J., Oster, G. & Nollert, P. (2003) Protein interactions and membrane geometry. *Biophys. J.* **84**, 854–868
46. Hofer, N., Aragao, D. & Caffrey, M. (2010) Crystallizing transmembrane peptides in lipidic mesophases. *Biophys. J.* **99**, L23–L25
47. Pye, V., Aragao, D., Lyons, J.A. & Caffrey, M. (2011) Overview of the 13th International Conference on the Crystallization of Biological Macromolecules. *Cryst. Growth Des.*, in the press
48. Forsythe, E., Achari, A. & Pusey, M.L. (2006) Trace fluorescent labeling for high-throughput crystallographyActa Crystallogr. *Sect. D Biol. Crystallogr.* **62**, 339–346
49. Kissick, D.J., Gualtieri, E.J., Simpson, G.J. & Cherezov, V. (2010) Nonlinear optical imaging of integral membrane proteins in lipidic mesophases. *Anal. Chem.* **82**, 491–497
50. Cherezov, V., Hanson, M.A., Griffith, M.T., Hilgart, M.C., Sanishvili, R., Nagarajan, V., Stepanov, S., Fischetti, R.F., Kuhn, P. & Stevens, R.C. (2009) Rastering strategy for screening and centring of microcrystal samples of human membrane proteins with a sub-10 microm size X-ray synchrotron beam. *J. R. Soc. Interface* **6**, S587–S597

Biochem. Soc. Symp. 78
Citation reference: Biochem. Soc. Trans. (2011) **39**, 733–740.

3

Sensing bilayer tension: bacterial mechanosensitive channels and their gating mechanisms

Ian R. Booth[1], Tim Rasmussen, Michelle D. Edwards, Susan Black, Akiko Rasmussen, Wendy Bartlett and Samantha Miller

Department of Medical Sciences, Institute of Medical Sciences, University of Aberdeen, Aberdeen AB25 2ZD, U.K.

Abstract

Mechanosensitive channels sense and respond to changes in bilayer tension. In many respects, this is a unique property: the changes in membrane tension gate the channel, leading to the transient formation of open non-selective pores. Pore diameter is also high for the bacterial channels studied, MscS and MscL. Consequently, in cells, gating has severe consequences for energetics and homoeostasis, since membrane depolarization and modification of cytoplasmic ionic composition is an immediate consequence. Protection against disruption of cellular integrity, which is the function of the major channels, provides a strong evolutionary rationale for possession of such disruptive channels. The elegant crystal structures for these channels has opened the way to detailed investigations that combine molecular genetics with electrophysiology and studies of cellular behaviour. In the present article, the focus is primarily on the structure of MscS, the small mechanosensitive channel. The description of the structure

[1]*To whom correspondence should be addressed (email i.r.booth@abdn.ac.uk).*

is accompanied by discussion of the major sites of channel–lipid interaction and reasoned, but limited, speculation on the potential mechanisms of tension sensing leading to gating.

Introduction

Mechanosensitive channels sense changes in the tension in the membrane bilayer [1]. In higher organisms the channels have diverse functions, but in bacteria, they are required for the maintenance of cell integrity during severe shifts in osmolarity, specifically during rapid transitions from high to low osmolarity (hypo-osmotic shock) [2]. Channel gating leads to the transient formation of high-conductance pores in the cytoplasmic membrane on a timescale of a few microseconds [3]. Detailed analyses of the permeability properties of all of the bacterial channels identified have yet to be completed. However, pores are large, probably ranging from 8 to 30 Å (1 Å = 0.1 nm) in diameter [4]. Consequently, the pores are predominantly non-selective, except on the basis of the size of the ion or organic solute that they pass. The best-studied channels are those encoded by the *mscL* gene (MscL) and the *yggB* gene (MscS) [5]. Crystal structures for the *Mycobacterium* MscL channel and the *Escherichia coli* MscS channel have been solved to ~3.5 Å and reveal the channels to be a homopentamer (MscL) and a homoheptamer (MscS) [6,7]. Most organisms possess both MscL and MscS, but there are often several additional MscS homologues, which form a large family of structurally related proteins [8]. Larger members of this family (>300 residues) bear a domain equivalent to MscS at their C-terminal end, but the sequences at the N-terminus are quite diverse [2]. Two of these MscS homologues from *E. coli*, YbdG and MscK, have been studied [9,10] and have been shown to be active, and, for both homologues, mutations in the 'MscS domain' modify the gating of the channel consistent with this proposed role (M.D. Edwards, U. Schumann, S. Black, S. Miller and I.R Booth, unpublished work). Different gene products form channels with different sensitivities to tension and each with a unique pore size. There is no evidence for the formation of hetero-oligomers between different gene products.

Most studies on mechanosensitive channels have used proteins from *E. coli* applying a range of techniques, but additional insights into MscL have been derived from investigations of the *Mycobacterium* and *Staphylococcus* channels [6,11–14]. Critical residues within the *E. coli* protein have been identified from mutagenic studies, and clues to the molecular motions have been derived from ESR using probes attached to cysteine residues introduced by site-directed mutagenesis [15–21]. Paradoxically, given the importance of tension sensing, considerably fewer studies have tried to understand how the proteins interact with the lipid bilayer [13,22–24]. In the present article, we principally seek to describe our understanding of the MscS channel. It is clear that MscL should share some fundamental similarities with regard to lipid–protein interactions, but it is structurally and mechanistically different and it is hazardous to extrapolate

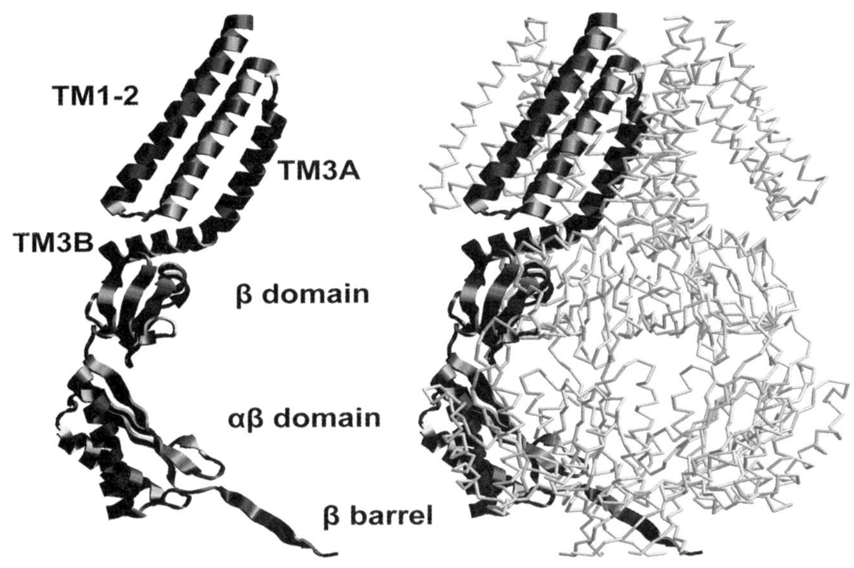

Figure 1. MscS structure (PDB code 2OAU)
The structure of a single subunit of MscS and the homoheptamer based on the structure solved by the Rees group [7]. The periplasmic surface is at the top.

too much from MscL to MscS. Moreover, MscS is a large and complex family of proteins exhibiting considerable diversity of structure, where insights into the differences in mechanisms between large and small homologues remain to be discovered. Work on MscL is described elsewhere in this issue of *Biochemical Society Transactions* [24a], and excellent reviews can be found in the literature [5].

The structure of MscS

Recognition that the *yggB* gene of *E. coli* encodes the MscS channel monomer paved the way to both the structure and to underpinning biochemical experiments to understand the structural changes during gating [2,7]. Rees and colleagues solved the structure for the *E. coli* MscS channel to ~3.5 Å [7]. MscS is a homoheptameric transmembrane protein with a large cytoplasmic domain. The membrane domain is composed of three α-helices per subunit, and the cytoplasmic domain is predominantly composed of β-sheets (Figure 1). Overall, however, each subunit (286 amino acids in *E. coli*) exhibits a complex domain structure. Both the N-terminal sequence (residues 1–26) and the last six residues of the C-terminal domain were not modelled in the crystal structure because of the lack of well-defined density corresponding to these sequences. However, models have been generated for the N-terminal region using EPR and molecular dynamics, and function is implied from the properties of mutants in this region [25].

Membrane domain

Each monomer contributes three helices (TM1–TM3) to the membrane domain. Density for TM1–TM2 (residues 27–91) was less well defined than for the main body of the protein [7,26]. They form a sensor that interacts with the lipids of the bilayer (see below) [7]. During gating, the TM1–TM2 domain appears to move as a rigid body that transmits tension to the pore via a short sequence (residues 92–95) [26]. In the original crystal structure, this region exists in extended chain conformation. Mutations in this sequence cause the channel to gate at lower bilayer tension [27,28]. The third helix, TM3, has two sections: residues 96–112 (TM3a) and 114–127 (TM3b). TM3a helices line the pore and are aligned at 25–35° to the membrane normal, which is also the pore axis. The residues lining the pore are hydrophobic. Two rings of leucine residues (Leu^{105} and Leu^{109}) form a pore constriction that must be expanded to generate the open pore. The packing surface between the adjacent TM3a helices is marked by a conserved alanine/glycine pattern that is a critical determinant of the gating tension [7,29]. TM3a ends with an Asn-Gly sequence that forms a bend leading to TM3b, which lies along the surface of the bilayer (see below).

Initially, there were some doubts regarding the nature of this form of MscS, i.e. whether it depicted the open or closed form of the channel [7,30]. A range of *in silico* analyses of the diameter of the pore indicated that the pore was very narrow and was too small to account for the conductance of the channel [30–32]. Additionally, the whole pore had been noted to be extremely hydrophobic, and molecular dynamics simulations led to the suggestion that the pore was closed by the inability to sustain a water column, the so-called 'vapour lock' [30,33]. A subsequent structure (3.45 Å resolution), derived from the crystallization of an MscS A106V mutant, trapped MscS in an conducting, open, state [26]. Here a pore approximately 13 Å in diameter was estimated using the HOLE program, compared with minimum 4.8 Å diameter estimated from the original structure of the wild-type protein [34,35]. The expanded pore diameter of the A106V structure is in the right range for the measured values of the conductance, and the expansion of the pore is of the magnitude that had previously been predicted for the closed-to-open transition [36].

TM3b

The role of TM3b in MscS is controversial and our understanding of the roles played by this sequence is still evolving [37–39]. Genetic analysis has led to considerable speculation regarding the role of the Asn-Gly hinge and TM3b, which is dealt with below. Asn^{112}, the final residue of TM3a, generates a helix break, which is essential for channel function, and creates a sharp bend around residue 113 [7]. Asn^{112} is essential for channel function, whereas amino acid substitutions at Gly^{113} are highly tolerated [37,40]. In both wild-type and A106V structures, a significant cavity exists where any side chain extending from position 113 would be placed. Indeed, a G113W mutant, which represents the largest amino acid substitution possible, forms a stable functional heptameric channel protein (M.D. Edwards, T. Rasmussen, S. Miller and I.R. Booth, unpublished work). The C-terminal end of TM3b has a pair of highly conserved basic residues (Arg^{128} and Arg^{131}). Mutation of either Arg^{128} or Arg^{131} to aspartate causes a slight loss of function in the channel [41]. These residues may have important roles in gating transitions through interactions with lipid headgroups.

The cytoplasmic domain (vestibule)

The vestibule structure is formed by the two predominantly β-sheet domains (Figure 1). The upper 'middle β-domain' is wholly β-sheet, and the seven subunits produce a relatively rigid cylindrical β-domain that surrounds the axis of the pore. It is connected to the lower '$\alpha\beta$-domain' by a short linker. The edges of the two domains, in combination with the connecting linker, define the perimeter of the lateral portals that connect the transmembrane pore to the cytoplasm [7]. During activation of the channels in the cell, considerable ion flux must take place through these portals to account for the rapid depletion of the cytoplasmic ion pools [34]. The vestibule inner surface is predominantly polar in character, but with few charged residues; consequently, the surface will not exhibit strong ion-binding properties. The lower domain is potentially more flexible than the upper domain due to the attachments to the 'middle β-domain' and the terminal β-barrel via extended peptide linkers [28]. Previously, we have referred to this as a Chinese lantern arrangement, indicating the potential for the vertical movement (in the direction of the membrane surface) that could lead to narrowing of the portals [42]. Various findings are consistent with the structure of this domain affecting the gating threshold for the channel, but the indirect nature of the assays makes explicit interpretation difficult.

The closed-to-open transition

The availability of two structures allows some prediction of the structural transition associated with gating. Thus, in the A106V structure the TM1–TM2 helices have rotated as a rigid body [26]. Although crystal structures are static representations of dynamic systems, comparison of the two available MscS structures does suggest significance of the TM1–TM2 sensor consistent with its ability to sense and respond to changes in its lipid context [7,26]. The movement of TM1–TM2 would result in a conformation of the linker to TM3a. What is clear from the more recent structure is that TM3a helices can tilt and separate [26].

In the closed structure, these helices achieve the very tight packing due to the match between surfaces populated with alanine (knobs) and glycine (grooves) residues [7,43]. In the A106V structure, the helices have separated, rotated and moved radially outwards and, in addition, they are aligned with the membrane normal [26]. It is this series of positional changes that generates the open pore by increasing the separation of the hydrophobic rings generated by Leu^{105} and Leu^{109}. The spatial separation of the TM3a helices is sustained by interactions at the base of TM3a between Ala^{110} and with Leu^{115} from TM3b of the adjacent subunit. At the upper end, there are two groups of interactions that could stabilize the open state: Ile^{97} with Ala^{98}/Ser^{95} of the next subunit and Thr^{93} with Gln^{92} of the adjacent subunit. The importance of these last interactions has not been tested in detail.

In conclusion, the structure of MscS provides a stable framework for creating a large pore in the cytoplasmic membrane in response to changes in bilayer tension. Rapid electrophysiological measurements have indicated that the transition from the closed to the open state can take place in a few microseconds [3]. Measurements made with the channel under constant moderate tension indicate that the channels open and close stochastically and have a potentially long open dwell time (>100 ms) [29]. At higher values of tension, alternative closed states may be attained, some of which are prevented from rapidly re-attaining the open state [38,44–46]. However, these observations must be tempered by the fact that, in cells, the open channel rapidly dissipates the transmembrane turgor pressure that is the physiological signal for channel activation. Thus the physiological significance of inactivated states *in vivo* is uncertain.

Lipid–protein interactions in MscS

Mechanosensitive channels gate in three specific circumstances. In cells, the channels gate when transmembrane turgor rises rapidly. This can be triggered by adapting cells to high osmolarity followed by rapid transfer into medium of low osmolarity [47,48]. Turgor increases due to the rapid inflow of water down the osmotic gradient generated by the high intracellular concentrations of osmotically active solutes accumulated by growing cells. The turgor pressure rises rapidly and has the potential to exceed 30 atm (≈3040 kPa) [49]. The gating of the channels allows the rapid release of the accumulated solutes, which reduces the driving force for water entry and therefore lowers turgor pressure. Clearly, the effects of turgor must be converted into variation of lipid bilayer tension such that the channels gate. In membrane patches, the channels can be triggered to gate either by exerting pressure on to the patch or by the insertion of amphipaths, such as lysophosphatidylcholine, into the lipid bilayer [50,51]. Pressure is exerted on to the patch by creation of negative pressure in the pipette to which the patch is attached, leading to an outwardly directed pressure

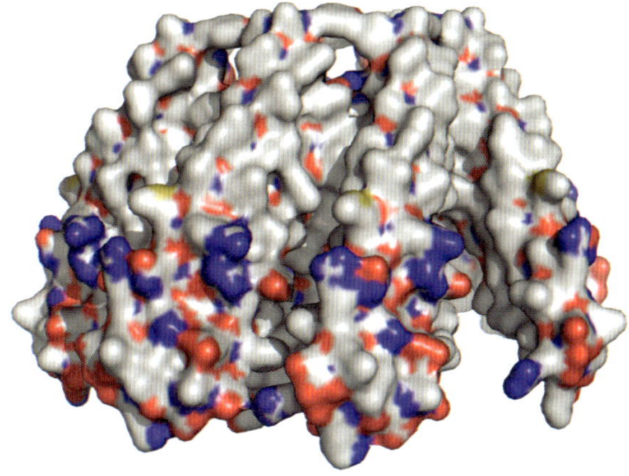

Figure 2. Surface representation of MscS membrane domain
Residues up to 113 are shown on the basis of the closed wild-type structure of *E. coli* MscS. Oxygen atoms are red, nitrogen atoms are blue, sulfur atoms are yellow, and carbon atoms are grey. The Figure was generated with PyMOL (http://www.pymol.org).

equivalent to turgor in cells. Amphipaths insert asymmetrically into the patch either by their inclusion in the pipette buffer (exposure to the outer leaflet of the membrane bilayer) or by inclusion in the bath. Their effects take some time to be seen, but generally result in lower pressure being required to gate the channels. In all three scenarios, the effects of pressure on the membrane must modify the lipid organization in proximity to the channel and the channel must possess a mechanism to sense the changes.

MscS has three principal lipid–protein interfaces. The TM1–TM2 helix pair was originally proposed to sense membrane tension [7]. It was noted from the crystal structure that TM1–TM2 acted as an independent domain and that its movements in the lipid bilayer would be transmitted to the pore helices through the linker between TM2 and TM3a. This may be the principal sensory domain of the MscS channel. The amino acid composition of TM1 and TM2 exhibits an extreme asymmetry. Low polarity is observed in the periplasmic half of both helices, and highly polar residues are found in the cytoplasmic half (Figure 2). This should predispose TM1–TM2 to interact differentially with the lipid headgroups at the periplasmic and cytoplasmic surfaces, which seems likely to be an intrinsic element of the lipid tension sensing. As indicated above, TM3b is predicted to lie along the inner membrane leaflet and presents a hydrophobic surface to the bilayer. The distal end (with respect to the pore) of TM3b has a pair of arginine residues that could interact with lipid headgroups and provide an anchor (Figure 3A) (see below). Finally, residues 1–26 were not modelled in the crystal structure because of a lack of density corresponding to this sequence [7]. Although this implies a degree of mobility in this region, structured models have been predicted [25]. Critically, mutations that replace Trp^{16} with other residues have been shown to modify MscS gating in response to pressure

[52]. Tryptophan residues are known to have a propensity to burrow into the lipid bilayer and to interact with the phospholipid headgroups. Substitution of tyrosine for tryptophan, which can also interact with lipid headgroups, increases the pressure required to gate the channel. Mutations that retain the hydrophobic character of tryptophan, but lack the ability to interact with lipid headgroups (phenylalanine or leucine) have more severe gating defects than the tryptophan-to-tyrosine change, consistent with lipid headgroup interactions at this position being critical to MscS function [52].

Genetic studies have explicitly address the interactions of TM1–TM2 with lipids [41,53]. Yoshimura and colleagues performed asparagine substitution mutagenesis in which the residues of TM1–TM2 were systematically replaced by asparagine [53]. The asparagine side chain has the capacity to hydrogen-bond to lipid headgroups and also with backbone carbonyls and amides of TM3a [54–56]. Yoshimura and colleagues noted position-specific changes in the gating pressure of the modified channels (Figure 3B) [53]. Asparagine positioned at the periphery of TM1 or TM2 causes the channel to become less sensitive to membrane tension. Introduction of two or more mutations at the equivalent positions relative to the bilayer causes complete loss of sensitivity to pressure. Thus F68N combined with A51N, where both mutations affect residues at the cytoplasmic end of the TM1–TM2 helix pair, was very insensitive to pressure and exhibited a greater change in pressure-sensitivity than either single mutant alone. In addition, when two mutations were introduced, but at opposite ends of the TM1–TM2 helix pair (e.g. F68N/I27N), there was no significant additive effect of the mutations on pressure sensitivity. An opposite effect was observed when the mutations were introduced towards the middle of the TM1–TM2 helix pair. Now the single mutations (e.g. I39N) caused the channel to gate at lower pressure. These data are supported by other studies in which mutations that increased gating were isolated [57]. Yoshimura and colleagues also confirmed that asparagine could be replaced by other charged residues (aspartate and lysine), but that the phenotype was not significantly altered by mutations that led to substitution of one hydrophobic residue for another [53]. Similar studies on MscL also identified that introduction of asparagine residues close to the periplasmic rim of TM1 and TM2 inactivated the channels by changing the tension sensitivity [58].

The original interpretation of these studies was that the hydrophobicity of the residue was important, but the studies did not provide a mechanistic interpretation of the mutations. The elucidation of the structures of many inner and outer membrane proteins has led to a general ‘rule’ about the location of tryptophan and tyrosine residues (and, to a lesser extent, asparagine and glutamine) [59,60]. In the majority of membrane proteins that have been studied, the tyrosine and tryptophan residues are found in positions proximal to the phospholipid headgroups. Both of the hydrophobic parts of the side chain tunnel down towards the fatty acid side chains of the lipids, whereas the polar groups form hydrogen bonds to oxygen atoms in the headgroups of the phospholipids. One of the most marked observations regarding both MscS and MscL was

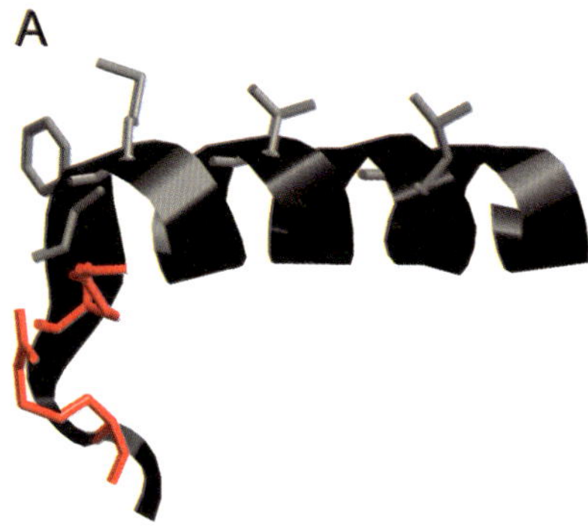

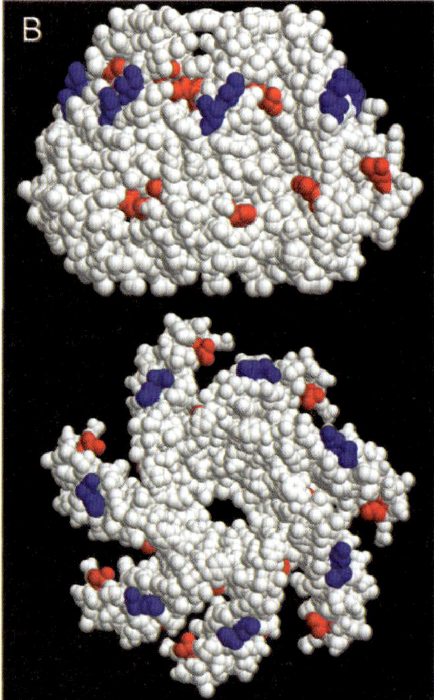

Figure 3. Lipid contact zones for the MscS channel
(**A**) Structure of TM3b. The structure is depicted in ribbon format, and the location of the hydrophobic amino acids side chains that are predicted to interact with the lipid bilayer are shown in black. The two highly conserved arginine residues (Arg^{128} and Arg^{131}) are shown in red. The orientation of these residues in the crystal structure may be explained by the presence of detergent (and/or lipid) in the crystals, but they could adopt the same orientation to interact with the lipid bilayer. (**B**) Positions of some of the asparagine-substitution mutants created by Yoshimura and colleagues in MscS [53]. Residues in red are strongly inhibitory, whereas those in blue activate MscS (gain-of-function). Note that in the membrane, TM1–TM2 may be turned slightly to pack against the outer surface of the TM3a helix. This has the effect of burying the 'red' residues against TM3a. In contrast, the blue residues always face the lipid bilayer whichever packing of TM1–TM2 with TM3a is preferred by the experimentalist.

the extremely limited numbers of tryptophan or tyrosine residues in the lipid exposed regions of the proteins. MscS has a single tyrosine (Tyr^{75}) and a single tryptophan (Trp^{16}) residue per subunit in membrane regions. Mutations at either of these positions modify channel gating, consistent with an interaction with the phospholipid headgroups [52]. However, the striking feature of the structures of TM1–TM2 in MscS is the extreme asymmetry in the distribution of polar amino acid residues at the periphery of these helices (Figure 2). This creates the potential for these helices to move within the bilayer in response to changes in membrane tension, causing the necessary changes in the linker TM2–TM3A that is required to open the pore. Thus one would argue that a fundamental aspect of mechanosensation for MscS (and also MscL) is the lack of strong interactions between the periphery of the TM1–TM2 sensor paddle with the lipid headgroups. This could allow for significant asymmetry in the movement in the TM1–TM2 domain in response to changes in lipid packing consequent to high turgor. However, it is still not clear how lipid packing is sensed. The structure of MscS implies a periplasmic pivoting of TM1–TM2 with the cytoplasmic ends describing an arc, which at its most extreme may involve TM1–TM2 passing across TM3b, as described in the A106V structure [26]. Mutant studies support this model, with tension being passed through the stretched protein strand that connects TM1–TM2 to TM3a, leading to the rotation and separation of TM3a helices. The observation that the introduction of asparagine (or charged amino acids) [53,57] into the middle region of these helices leads to gating at low applied pressure (gain-of-function) is also consistent with this model. The requirement for hydrogen-bond formation (for the minimum energy state of the protein) between asparagine (or aspartate or lysine) is satisfied via interaction with the phospholipid headgroups, which requires a positional change in the TM1–TM2 equivalent to the pivot arising from changes in lipid packing. Blount and colleagues also identified that introduction of polar residues (V40D or V40K mutations) in the middle of TM1 caused MscS to gate at lower tensions, consistent with the formation of TM1–lipid headgroup interactions, leading to repositioning of TM1–TM2 towards the ‘open state’ [57]. We propose that this permanent change in orientation is equivalent to the first part of the molecular motion associated with gating of the wild-type channel, hence less energy is required from membrane tension to achieve the open state. The loss-of-function asparagine mutants could arise by anchoring TM1–TM2 to lipid headgroups due to the peripheral location of the inhibitory mutations. Alternatively, these asparagine residues may form hydrogen bonds to the peptide backbone of TM3a, thereby increasing the energy needed to allow the pivoting of TM1–TM2 and gate the channel.

A second study investigated the roles of the highly conserved basic residues (Arg^{128} and Arg^{131}) at the C-terminal end of TM3b [41]. Mutation of either Arg^{128} or Arg^{131} to aspartate causes a slight loss of function in the channel. The phenotype of the arginine-to-aspartate mutants has been interpreted in terms of breaking a salt bridge interaction with acidic residues at the base of the

TM1–TM2 sensor paddle (Asp^{62}) [61]. Such a role is problematic in that crystal structures suggest these residues to be at least 5 Å apart, which would provide only for a very weak interaction. Molecular dynamics simulations have suggested that TM1–TM2 is packed tightly against the TM3a pore when the protein is inserted in the lipid bilayer, which would only exacerbate the problems of making a salt bridge between Arg^{128}/Arg^{131} and Asp^{62} in TM1–TM2. These residues may have important roles in gating transitions through interactions with lipid headgroups. In the symmetrical homoheptamer, these arginine residues form a ring of 14 positive charges that can interact with the phospholipid headgroups and thereby anchor the extremities of TM3b. During the formation of the open state, there is rotation of TM3a and this 'turning force' must be absorbed by TM3b to attain stability in the open state. The hydrophobic residues of TM3b are likely to be anchored in the bilayer lipid phase, whereas the interactions of Arg^{128} and Arg^{131} with lipid headgroups would provide further resistance to rotation of TM3b. Hence any tendency of TM3b to rotate would be resisted, leading to stabilization of the open state. Support for this analysis comes from insertion of helix-breaking residues into TM3b close to the pore axis [37,46]. These mutant channels almost invariably display instability of the open state consistent with loss of resistance within TM3b. Recent work that has deployed selection for increased gating at low tension to identify mutations from a random genetic pool supports the potential importance of these residues in stabilizing the closed state [39].

The larger members of the MscS family pose an intriguing aspect to the problem of protein–lipid interactions and tension sensing [10]. The majority of MscS homologues possess multiple additional N-terminal transmembrane helices. These proteins can be broadly divided into two classes: those that have an additional two helices per monomer and those that have eight extra helices per subunit (Figure 4). Surrounding the core MscS domain with a number of transmembrane helices generates two potential problems. First, even for the homologues with just two extra helices per subunit, there is the possibility of a ring of protein that isolates the TM1–TM2 sensor from the lipid bilayer. Secondly, it raises the possibility that, in the larger channels, the sensor is not the helices immediately adjacent to TM3a, but is more remote and requires a more complex mechanism for transmission of the tension to the pore. Studying these proteins offers considerable rewards for the understanding of the mechanism of tension sensing by mechanosensitive channels.

In conclusion, major progress has been made on many aspects of the mechanism of pore formation by mechanosensitive channels. However, the central conundrum of lipid–protein interactions and tension sensing has not yet been solved. It is an important problem, since this class of membrane protein has potentially the most dynamic set of interactions with the lipid bilayer. Their mechanism involves sensing changes in tension, which probably reflect changes in lipid packing or organization. Although membrane proteins do sometimes display changes in activity in the context of changing the lipid composition of

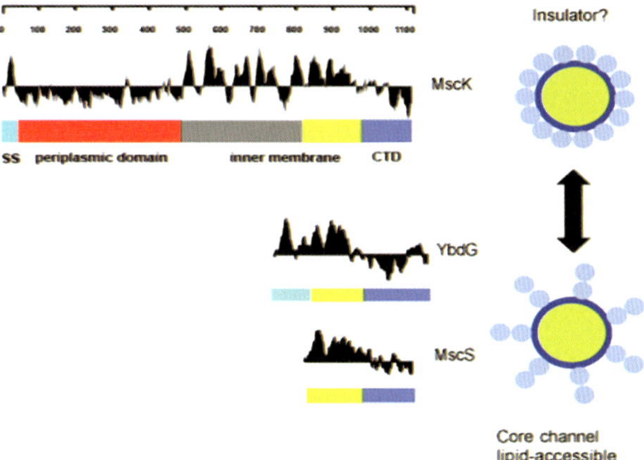

Figure 4. Membrane domains of MscS homologues
The hydrophobicity plots for MscS, MscK and YbdG are shown with a cartoon for MscS in which the membrane domain is yellow and the cytoplasmic vestibule is blue. Bars of the same colour are drawn beneath the hydrophobicity plot to indicate the corresponding sequence segments. Additional domains are coloured grey, red and pale blue (MscK, inner membrane, periplasmic and signal sequences respectively) and pale blue (YbdG, additional inner membrane domains). The potential arrangements of the additional transmembrane helices are shown with either annular or radial arrangements of helices that can leave the central channel domain either insulated (above) or accessible to lipid (below) respectively. SS, signal sequence; CTD, C-terminal domain.

the membrane, in particular there is selectivity for lipid headgroups, there is little evidence for this being a dynamic property of the membrane (although for a dissenting view, see [62]). This makes mechanosensitive channels almost unique among membrane proteins since the lipid changes from being a 'passive' context in which the protein functions, to an active component that by its own modulation determines protein activity.

Acknowledgements

We thank our collaborators Jim Naismith, Wenjiang Wang, Stuart Conway, Tony Lee, Jim Bowie and Paul Blount and our former colleagues, Ulrike Schumann and Chan Li, for their informed contributions to our work.

Funding

The Wellcome Trust [grant numbers GR077564MA and WT09255-2MA] has enabled our work on MscS and related channels for 15 years and we are extremely grateful for their financial support.

References

1. Sukharev, S.I., Sigurdson, W.J., Kung, C. & Sachs, F. (1999) Energetic and spatial parameters for gating of the bacterial large conductance mechanosensitive channel, MscL. *J. Gen. Physiol.* **113**, 525–540

2. Levina, N., Totemeyer, S., Stokes, N.R., Louis, P., Jones, M.A. & Booth, I.R. (1999) Protection of *Escherichia coli* cells against extreme turgor by activation of MscS and MscL mechanosensitive channels: identification of genes required for MscS activity. *EMBO J.* **18**, 1730–1737
3. Shapovalov, G. & Lester, H.A. (2004) Gating transitions in bacterial ion channels measured at 3 μs resolution. *J. Gen. Physiol.* **124**, 151–161
4. Cruickshank, C.C., Minchin, R.F., Le Dain, A.C. & Martinac, B. (1997) Estimation of the pore size of the large-conductance mechanosensitive ion channel of *Escherichia coli*. *Biophys. J.* **73**, 1925–1931
5. Kung, C., Martinac, B. & Sukharev, S. (2010) Mechanosensitive channels in microbes. *Annu. Rev. Microbiol.* **64**, 313–329
6. Chang, G., Spencer, R.H., Lee, A.T., Barclay, M.T. & Rees, D.C. (1998) Structure of the MscL homolog from *Mycobacterium tuberculosis*: a gated mechanosensitive ion channel. *Science* **282**, 2220–2226
7. Bass, R.B., Strop, P., Barclay, M. & Rees, D.C. (2002) Crystal structure of *Escherichia coli* MscS, a voltage-modulated and mechanosensitive channel. *Science* **298**, 1582–1587
8. Pivetti, C.D., Yen, R., Miller, S., Busch, W., Tseng, Y.H., Booth, I.R. & Saier, M.H. (2003) Two families of mechanosensitive channel proteins. *Microbiol. Mol. Biol. Rev.* **67**, 66–85
9. Li, Y., Moe, P.C., Chandrasekaran, S., Booth, I.R. & Blount, P. (2002) Ionic regulation of MscK, a mechanosensitive channel from *Escherichia coli*. *EMBO J.* **21**, 5323–5330
10. Schumann, U., Edwards, M.D., Rasmussen, T., Bartlett, W., van West, P. & Booth, I.R. (2010) YbdG in *Escherichia coli* is a threshold-setting mechanosensitive channel with MscM activity. *Proc. Natl. Acad. Sci. U.S.A.* **107**, 12664–12669
11. Liu, Z., Gandhi, C.S. & Rees, D.C. (2009) Structure of a tetrameric MscL in an expanded intermediate state. *Nature* **461**, 120–124
12. Maurer, J.A., Elmore, D.E., Lester, H.A. & Dougherty, D.A. (2000) Comparing and contrasting *Escherichia coli* and *Mycobacterium tuberculosis* mechanosensitive channels (MscL): new gain of function mutations in the loop region. *J. Biol. Chem.* **275**, 22238–22244
13. Powl, A.M., East, J.M. & Lee, A.G. (2005) Heterogeneity in the binding of lipid molecules to the surface of a membrane protein: hot spots for anionic lipids on the mechanosensitive channel of large conductance MscL and effects on conformation. *Biochemistry* **44**, 5873–5883
14. Yoshimura, K., Batiza, A., Schroeder, M., Blount, P. & Kung, C. (1999) Hydrophilicity of a single residue within MscL correlates with increased channel mechanosensitivity. *Biophys. J.* **77**, 1960–1972
15. Blount, P., Sukharev, S.I., Moe, P.C., Nagle, S.K. & Kung, C. (1996) Towards an understanding of the structural and functional properties of MscL, a mechanosensitive channel in bacteria. *Biol. Cell* **87**, 1–8
16. Blount, P., Sukharev, S.I., Schroeder, M.J., Nagle, S.K. & Kung, C. (1996) Single residue substitutions that change the gating properties of a mechanosensitive channel in *Escherichia coli*. *Proc. Natl. Acad. Sci. U.S.A* **93**, 11652–11657
17. Blount, P., Schroeder, M.J. & Kung, C. (1997) Mutations in a bacterial mechanosensitive channel change the cellular response to osmotic stress. *J. Biol. Chem.* **272**, 32150–32157
18. Moe, P.C., Blount, P. & Kung, C. (1998) Functional and structural conservation in the mechanosensitive channel MscL implicates elements crucial for mechanosensation. *Mol. Microbiol.* **28**, 583–592
19. Ou, X., Blount, P., Hoffman, R.J. & Kung, C. (1998) One face of a transmembrane helix is crucial in mechanosensitive channel gating. *Proc. Natl. Acad. Sci. U.S.A.* **95**, 1471–11475
20. Moe, P.C., Levin, G. & Blount, P. (2000) Correlating a protein structure with function of a bacterial mechanosensitive channel. *J. Biol. Chem.* **275**, 31121–31127
21. Iscla, I., Levin, G., Wray, R., Reynolds, R. & Blount, P. (2004) Defining the physical gate of a mechanosensitive channel, MscL, by engineering metal-binding sites. *Biophys. J.* **87**, 3172–3180
22. Moe, P. & Blount, P. (2005) Assessment of potential stimuli for mechano-dependent gating of MscL: effects of pressure, tension, and lipid headgroups. *Biochemistry* **44**, 12239–12244

23. Powl, A.M., East, J.M. & Lee, A.G. (2003) Lipid–protein interactions studied by introduction of a tryptophan residue: the mechanosensitive channel MscL. *Biochemistry* **42**, 14306–14317
24. Powl, A.M., Wright, J.N., East, J.M. & Lee, A.G. (2005) Identification of the hydrophobic thickness of a membrane protein using fluorescence spectroscopy: studies with the mechanosensitive channel MscL. *Biochemistry* **44**, 5713–5721
24a. Lee, A.G. (2011) Lipid–protein interactions. *Biochem. Soc. Trans.* **39**, 761–766
25. Vasquez, V., Sotomayor, M., Cortes, D.M., Roux, B., Sculten, K. & Perozo, E. (2008) Three-dimensional architecture of membrane-embedded MscS in the closed conformation. *J. Mol. Biol.* **378**, 55–70
26. Wang, W., Black, S.S., Edwards, M.D., Miller, S., Morrison, E.L., Bartlett, W., Dong, C., Naismith, J. & Booth, I.R. (2008) The structure of an open form of an *E. coli* mechanosensitive channel at 3.45 Å resolution. *Science* **321**, 1179–1183
27. Li, C., Edwards, M.D., Jeong, H., Roth, J. & Booth, I.R. (2007) Identification of mutations that alter the gating of the *Escherichia coli* mechanosensitive channel protein, MscK. *Mol. Microbiol.* **64**, 560–574
28. Miller, S., Bartlett, W., Chandrasekaran, S., Simpson, S., Edwards, M.D. & Booth, I.R (2003) Domain organization of the MscS mechanosensitive channel of *Escherichia coli*. *EMBO J.* **22**, 36–46
29. Edwards, M.D., Li, Y., Kim, S., Miller, S., Bartlett, W., Black, S., Dennison, S., Iscla, I., Blount, P., Bowie, J.U. & Booth, I.R. (2005) Pivotal role of the glycine-rich TM3 helix in gating the MscS mechanosensitive channel. *Nat. Struct. Mol. Biol.* **12**, 113–119
30. Anishkin, A. & Sukharev, S. (2004) Water dynamics and dewetting transition in the small mechanosensitive channel MscS. *Biophys. J.* **86**, 2883–2895
31. Sotomayor, M., Van Der Straaten, T.A., Ravaioli, U. & Schulten, K. (2006) Electrostatic properties of the mechanosensitive channel of small conductance MscS. *Biophys. J.* **90**, 3496–3510
32. Sotomayor, M., Vasquez, V., Perozo, E. & Schulten, K. (2006) Ion conduction through MscS as determined by electrophysiology and simulation. *Biophys. J.* **92**, 886–902
33. Beckstein, O. & Sansom, M.S. (2004) The influence of geometry, surface character, and flexibility on the permeation of ions and water through biological pores. *Phys. Biol.* **1**, 42–52
34. Steinbacher, S., Bass, R., Strop, P. & Rees, D.C. (2007) Structures of the prokaryotic mechanosensitive channels MscL and MscS. *Curr. Top. Membr.* **58**, 1–24
35. Vora, T., Corry, B. & Chung, S.H. (2006) Brownian dynamics investigation into the conductance state of the MscS channel crystal structure. *Biochim. Biophys. Acta* **1758**, 730–737
36. Sukharev, S. (2002) Purification of the small mechanosensitive channel of *Escherichia coli* (MscS): the subunit structure, conduction, and gating characteristics in liposomes. *Biophys. J.* **83**, 290–298
37. Akitake, B., Anishkin, A., Liu, N. & Sukharev, S. (2007) Straightening and sequential buckling of the pore-lining helices define the gating cycle of MscS. *Nat. Struct. Mol. Biol.* **14**, 1141–1149
38. Belyy, V., Kamaraju, K., Akitake, B., Anishkin, A. & Sukharev, S. (2010) Adaptive behavior of bacterial mechanosensitive channels is coupled to membrane mechanics. *J. Gen. Physiol.* **135**, 641–652
39. Koprowski, P., Grajkowski, W., Isacoff, E.Y. & Kubalski, A. (2011) Genetic screen for potassium leaky small mechanosensitive channels (MscS) in *Escherichia coli*: recognition of cytoplasmic β domain as a new gating element. *J. Biol. Chem.* **286**, 877–888
40. Edwards, M.D., Bartlett, W. & Booth, I.R. (2008) Pore mutations of the *Escherichia coli* MscS channel affect desensitization but not ionic preference. *Biophys. J.* **94**, 3003–3013
41. Nomura, T., Sokabe, M. & Yoshimura, K. (2008) Interaction between the cytoplasmic and transmembrane domains of the mechanosensitive channel MscS. *Biophys. J.* **94**, 1638–1645
42. Booth, I.R. & Louis, P. (1999) Managing hypoosmotic stress: aquaporins and mechanosensitive channels in *Escherichia coli*. *Curr. Opin. Microbiol.* **2**, 166–169
43. Kim, S., Chamberlain, A.K. & Bowie, J.U. (2004) Membrane channel structure of *Helicobacter pylori* vacuolating toxin: role of multiple GXXXG motifs in cylindrical channels. *Proc. Natl. Acad. Sci. U.S.A.* **101**, 5988–5991

44. Akitake, B., Anishkin, A. & Sukharev, S. (2005) The "dashpot" mechanism of stretch-dependent gating in MscS. *J. Gen. Physiol.* **125**, 143–154
45. Anishkin, A., Akitake, B. & Sukharev, S. (2008) Characterization of the resting MscS: modeling and analysis of the closed bacterial mechano-sensitive channel of small conductance. *Biophys. J.* **94**, 1252–1266
46. Belyy, V., Anishkin, A., Kamaraju, K., Liu, N. & Sukharev, S. (2010) The tension-transmitting 'clutch' in the mechanosensitive channel MscS. *Nat. Struct. Mol. Biol.* **17**, 451–458
47. Berrier, C., Coulombe, A., Szabo, I., Zoratti, M. & Ghazi, A. (1992) Gadolinium ion inhibits loss of metabolites induced by osmotic shock and large stretch-activated channels in bacteria. *Eur. J. Biochem.* **206**, 559–565
48. Schleyer, M., Schmid, R. & Bakker, E.P. (1993) Transient, specific and extremely rapid release of osmolytes from growing cells of *Escherichia coli* K-12 exposed to hypoosmotic shock. *Arch. Microbiol.* **160**, 424–431
49. Kung, C. (2005) A possible unifying principle for mechanosensation. *Nature* **436**, 647–654
50. Martinac, B., Adler, J. & Kung, C. (1990) Mechanosensitive ion channels of *E. coli* activated by amphipaths. *Nature* **348**, 261–263
51. Martinac, B., Buehner, M., Delcour, A.H., Adler, J. & Kung, C. (1987) Pressure-sensitive ion channel in *Escherichia coli*. *Proc. Natl. Acad. Sci. U.S.A.* **84**, 2297–2301
52. Rasmussen, A., Rasmussen, T., Edwards, M.D., Schauer, D., Schumann, U., Miller, S. & Booth, I.R. (2007) The role of tryptophan residues in the function and stability of the mechanosensitive channel MscS from *Escherichia coli*. *Biochemistry* **46**, 10899–10908
53. Nomura, T., Sokabe, M. & Yoshimura, K. (2006) Lipid–protein interaction of the MscS mechanosensitive channel examined by scanning mutagenesis. *Biophys. J.* **91**, 2874–2881
54. Adamian, L. & Liang, J. (2002) Interhelical hydrogen bonds and spatial motifs in membrane proteins: polar clamps and serine zippers. *Proteins* **47**, 209–218
55. Adamian, L., Jackups, Jr, R., Binkowski, T.A. & Liang, J. (2003) Higher-order interhelical spatial interactions in membrane proteins. *J. Mol. Biol.* **327**, 251–272
56. Lear, J.D., Gratkowski, H., Adamian, L., Liang, J. & DeGrado, W.F. (2003) Position-dependence of stabilizing polar interactions of asparagine in transmembrane helical bundles. *Biochemistry* **42**, 6400–6407
57. Okada, K., Moe, P.C. & Blount, P. (2002) Functional design of bacterial mechanosensitive channels: comparisons and contrasts illuminated by random mutagenesis. *J. Biol. Chem.* **277**, 27682–27688
58. Yoshimura, K., Nomura, T. & Sokabe, M. (2004) Loss-of-function mutations at the rim of the funnel of mechanosensitive channel MscL. *Biophys. J.* **86**, 2113–2120
59. White, S.H. & Wimley, W.C. (1999) Membrane protein folding and stability: physical principles. *Annu. Rev. Biophys. Biomol. Struct.* **28**, 319–365
60. Killian, J.A. & von Heijne, G. (2000) How proteins adapt to a membrane–water interface. *Trends Biochem. Sci.* **25**, 429–434
61. Sotomayor, M. & Schulten, K. (2004) Molecular dynamics study of gating in the mechanosensitive channel of small conductance MscS. *Biophys. J.* **87**, 3050–3065
62. Romantsov, T., Battle, A.R., Hendel, J.L., Martinac, B. & Wood, J.M. (2010) Protein localization in *Escherichia coli* cells: comparison of the cytoplasmic membrane proteins ProP, LacY, ProW, AqpZ, MscS, and MscL. *J. Bacteriol.* **192**, 912–924

Biochem. Soc. Symp. 78
Citation reference: Biochem. Soc. Trans. (2011) **39**, 741–745.

4

The pumps that fuel a sperm's journey

Michael Jakob Clausen, Poul Nissen and Hanne Poulsen[1]

PUMPKIN – Centre for Membrane Pumps in Cells and Disease, Danish National Research Foundation, Gustav Wieds Vej 10C, DK-8000, Aarhus C, Denmark, and Department of Molecular Biology, Aarhus University, Gustav Wieds Vej 10C, DK-8000, Aarhus C, Denmark

Abstract

The sole purpose of a sperm cell is to carry genetic information from a male to a female egg. In order to accomplish this quest, the sperm cell must travel a long distance through a constantly changing environment. The success of this journey depends on membrane proteins that are uniquely expressed in sperm cells. One of these proteins is the $\alpha 4$ isoform of the sodium pump. This pump is optimized to cope with the ionic environment characteristic of the female reproductive tract, and its activity may be tightly coupled with secondary transporters that maintain cytoplasmic pH. Pharmacological inhibition of $\alpha 4$ is sufficient to inhibit sperm motility, and significant differences around the inhibitor-binding site compared with the ubiquitous $\alpha 1$ isoform, make $\alpha 4$ a feasible target in rational drug development.

Introduction

Of all the cell types in animals, none will venture further than spermatozoa. After their morphological differentiation, they reside in the caudal part of the epididymis of the male gonads, where they are kept in a quiescent state

[1]*To whom correspondence should be addressed (email hp@mb.au.dk).*

until their release. Upon ejaculation, spermatozoa are confronted with a hostile and constantly changing extracellular environment throughout the female reproductive tract. To cope with the special circumstances, spermatozoa express a unique variety of transporters, channels and pumps that enable the cells to utilize the strange environment for their maturation. One of the distinctive membrane components is the sodium pump isoform $\alpha 4$, which is only found in spermatozoa. In the present paper, we summarize relevant $\alpha 4$ data in the context of sperm development to approach an answer to the fundamental question: why do spermatozoa need a unique sodium pump isoform?

The sodium pump and the α-subunit isoforms

The reaction scheme of the Na^+/K^+-ATPase is a cycle where three Na^+ are excluded from the cell in exchange for two K^+. This reaction is against the chemical gradients of both Na^+ and K^+, and for Na^+ also against the electrical potential. The reaction therefore depends on energy, which comes from the hydrolysis of ATP.

The complete sodium pump is a heteromeric complex with two essential subunits (α and β) and a third being regulatory (FXYD or γ). The catalytic α-subunit is responsible for ion transport and ATP hydrolysis, and it is the target of the specific sodium pump inhibitor ouabain. The other subunit is the β-subunit, which is required for the correct processing of the complex in the Golgi apparatus and for its transport to the membrane [1,2].

In mammals, four different Na^+/K^+-ATPase α-subunit isoenzymes have been identified, each with characteristic cell-type expression profiles and kinetic properties. The four α-subunits are systematically named $\alpha 1$–$\alpha 4$. β-Subunits have been found in three different forms known as $\beta 1$, $\beta 2$ and $\beta 3$. Pairwise comparison of the four α-subunit isoforms from three different mammals reveals a high degree of conservation across species for $\alpha 1$–$\alpha 3$, whereas $\alpha 4$ is less preserved (Figure 1). The somewhat lower conservation of the $\alpha 4$ isoforms between species could reflect the fact that $\alpha 4$ serves the simple functional role of transporting ions, while the complex interactions in larger protein networks that have been described for the other isoforms might be less important in the sperm. Alternatively, the $\alpha 4$ variability may reflect the high degree of species specialization of spermatozoa, a cell type that diverges significantly between otherwise closely related species.

The sperm-specific pump

The $\alpha 4$ isoform of the sodium pump was identified in 1987 in a screen of genomic DNA from human leucocytes using probes derived from the sheep $\alpha 1$ and rat $\alpha 2$ sodium pumps [3]. Later, Northern blot analysis with total RNA from different human tissues detected transcripts of the $\alpha 4$ isoform in testis exclusively [4]. The testes-specific sodium pump isoform has now been detected both at mRNA and

		1	2	3	4	5	6	7	8	9	10	11	12
Homo sapiens $\alpha 1$	1		98.63	96.77	86.91	87.40	86.82	86.41	86.51	86.22	77.81	78.32	76.60
Sus scrofa $\alpha 1$	2	98.63		97.17	86.91	87.40	86.82	86.61	86.61	86.41	77.42	78.41	76.31
Rattus norvegicus $\alpha 1$	3	96.77	97.17		86.33	86.72	86.52	85.63	85.63	85.53	77.52	78.41	76.60
Homo sapiens $\alpha 2$	4	86.91	86.91	86.33		99.12	99.12	86.52	86.72	86.33	78.37	78.88	77.25
Sus scrofa $\alpha 2$	5	87.40	87.40	86.72	99.12		98.63	86.72	86.91	86.52	78.27	79.07	77.54
Rattus norvegicus $\alpha 2$	6	86.82	86.82	86.52	99.12	98.63		86.52	86.72	86.33	78.37	78.97	77.44
Homo sapiens $\alpha 3$	7	86.41	86.61	85.63	86.52	86.72	86.52		99.51	99.51	76.84	77.15	75.34
Sus scrofa $\alpha 3$	8	86.51	86.61	85.63	86.72	86.91	86.72	99.51		99.21	76.94	77.15	75.34
Rattus norvegicus $\alpha 3$	9	86.22	86.41	85.53	86.33	86.52	86.33	99.51	99.21		76.65	77.06	75.24
Homo sapiens $\alpha 4$	10	77.81	77.42	77.52	78.37	78.27	78.37	76.84	76.94	76.65		86.12	82.85
Sus scrofa $\alpha 4$	11	78.32	78.41	78.41	78.88	79.07	78.97	77.15	77.15	77.06	86.12		81.59
Rattus norvegicus $\alpha 4$	12	76.60	76.31	76.60	77.25	77.54	77.44	75.34	75.34	75.24	82.85	81.59	

Figure 1. Conservation of the four sodium pump isoforms from three different mammalian species

Pairwise comparisons of the four different sodium pump α subunit isoforms from human (*Homo sapiens*), pig (*Sus scrofa*) and rat (*Rattus norvegicus*). All numbers are percentage identities.

protein levels in several mammalian species, and, at a cellular level, its expression seems to be confined to gametes [5–7]. In both humans and mice, the gene encoding $\alpha 4$ is found on chromosome 1 in close proximity to the gene encoding $\alpha 2$ [8,9], indicating that $\alpha 4$ could originate from a gene duplication of $\alpha 2$, in line with the sequence identity data showing that $\alpha 4$ has slightly higher identity with $\alpha 2$ than with the others (Figure 1).

Expression of $\alpha 4$ during spermatogenesis

Spermatogenesis is the differentiation of spermatogonia into spermatozoa. This process takes place in the seminiferous tubules where primary spermatocytes are derived from diploid spermatogonia. Each primary spermatocyte is then divided into two secondary spermatocytes, and they each give rise to two immature spermatids. Finally, the spermatids develop into the morphologically characteristic spermatozoa that travel from the seminiferous tubules to the epididymides.

Expression of $\alpha 4$ is driven by the transcription factor CREM [CRE (cAMP-responsive element) modulator] τ, which recognizes a CRE motif located 263 bp upstream of the transcription initiation site [10]. Several isoforms of CREM are expressed in male germ cells, both the antagonistic forms α, β and γ and the activator form τ. During spermatogenesis, the levels of antagonistic CREM forms decrease from early primary spermatocytes until the late spermatocyte phase, when the levels of stimulatory CREMτ increase significantly [11]. This correlates with the reported $\alpha 4$ mRNA levels in rat, which increase during the spermatocyte phase, and protein levels rise dramatically during the spermatid phase to peak in spermatozoa [12].

The production of cAMP leads to activation of protein kinase A and downstream phosphorylation events, and it has long been recognized that the hypermobility of sperm depends on cAMP production [13,14]. There are a

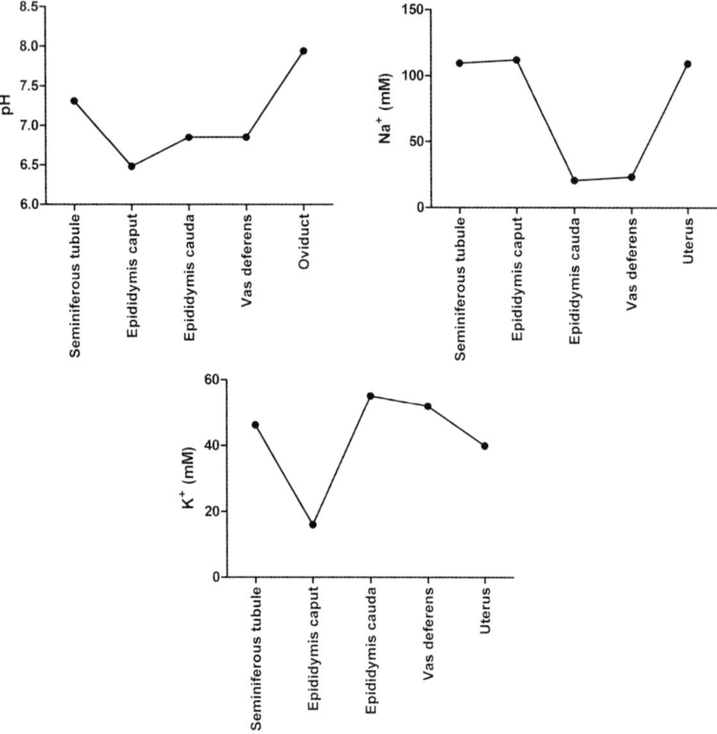

Figure 2. The ionic environment through which sperm cells travel
Alterations in the ionic environment of rodent reproductive tracts. All values are from [31], except for uterus ion concentrations, which are from [32], and oviduct pH, which is from [33].

number of adenylate cyclases expressed in sperm, but a critical pathway is the bicarbonate-activated soluble adenylate cyclase. Male mice lacking this adenylate cyclase are infertile, and they produce spermatozoa with severely impaired mobility [15]. Numerous processes in the spermatozoa are switched on by cAMP; one is the expression of $\alpha 4$.

Spermatozoa face a changing extracellular environment

When residing in the cauda of the epididymis, the mature sperm cells are in a quiescent state to preserve energy reserves and minimize endogenous oxidizing agents resulting from mitochondrial activity. One of the factors known to keep unejaculated sperm cells in the dormant state is the internal pH, which is low due to the acidic pH in the caudal epididymis [16,17]. During the passage through the female reproductive tract, the pH of the environment increases towards 8 (Figure 2), and alkalinization of the spermatozoon's cytoplasm is one of the important factors leading to hyperactivation of motility and the acrosome reaction.

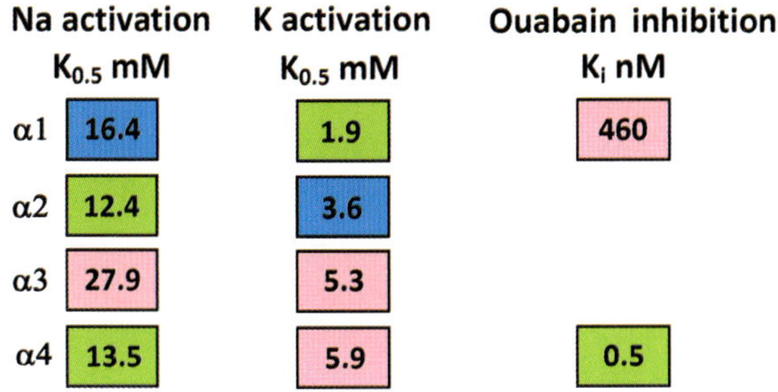

Figure 3. Ion and ouabain affinities for the four sodium pump isoforms
Apparent ion affinities of the four rat sodium pump isoforms expressed in Sf9 insect cells. Sodium affinity was determined in the presence of 30 mM potassium; potassium affinity was determined in the presence of 120 mM sodium [18]. Ouabain K_i values are from human sperm homogenates [20]. Green, blue and red indicate high, medium and low affinities respectively. Values taken from [18].

To get a better idea of why spermatozoa are equipped with a unique sodium pump isoform, it is important to consider the ion affinities of the different isoforms compared with their ionic environments.

Different systems have been used to characterize the enzymatic properties of the four sodium pump isoforms. Comparison of apparent ion affinities between all four isoforms has so far only been done in membrane fractions from Sf9 insect cells stably transfected with the different sodium pump isoforms. The ouabain-sensitive ATPase activity was determined from the release of $^{32}P_i$ under various ion concentrations. In this system, the apparent affinities for sodium differ between the four isoforms, with $\alpha3$ having comparatively low affinity, $\alpha2$ and $\alpha4$ having high affinities, and $\alpha1$ in between. The pattern of potassium affinity is different. Here, the lowest affinity is seen for $\alpha3$ and $\alpha4$, the highest for $\alpha1$, and $\alpha2$ is in between (Figure 3) [18]. Interestingly, spermatozoa express the two isoforms that differ the most in potassium affinity.

In the cauda of the epididymis where spermatozoa are in the quiescent state, the concentrations of sodium and potassium are exceptionally low and high respectively, which is optimal for sodium pump activity. Through the female reproductive tract, the concentration of sodium increases, but the potassium levels remain sufficiently high for $\alpha4$ to be fully active (Figure 2), so $\alpha4$'s relatively low potassium affinity may reflect an adaptation to the environment.

Sperm motility depends on $\alpha4$

In rodents, the $\alpha1$ isoform of the sodium pump has an exceptionally low sensitivity to ouabain and is only completely inhibited by concentrations approaching 10 mM. In contrast, the $\alpha4$ subunit is highly sensitive to ouabain and is completely inhibited at concentrations of ~1 μM. Such substantial differences offer the opportunity to inhibit $\alpha4$ specifically with low ouabain concentrations and both pumps with high concentrations. Using this approach, it has been demonstrated that inhibition of $\alpha4$ alone is sufficient to severely impair sperm motility, and additional inhibition of $\alpha1$ had no additive effect [19]. The same characteristic is seen in human sperm cells, although the ouabain-sensitivity difference window is smaller [20]. Recently, male $\alpha4$-null mice were shown to be completely sterile, and the spermatozoa's motility was severely impaired. This new mouse model has confirmed that the activity of $\alpha4$ is essential for the dedicated motility of spermatozoa, whereas $\alpha1$ may serve the role of a housekeeping enzyme [21].

A study of transgenic mice with protamine 1 promoter-driven expression of EGFP (enhanced green fluorescent protein)-tagged rat $\alpha4$ has previously cast light on the relationship between spermatozoa motility, ion transport and transmembrane potential [21]. The innate role of the protamine 1 promoter is to direct expression of protamine 1 in round spermatids [22], and spermatozoa of the transgenic mice express rat $\alpha4$–EGFP throughout the flagellum. The high ouabain sensitivity of $\alpha4$ was used to determine the surface levels of the pump from binding fluorescently labelled ouabain, demonstrating increased levels of $\alpha4$ in the transgenic mouse compared with wild-type.

The overexpression of the rat $\alpha4$–EGFP construct led to higher motility parameters and to membrane hyperpolarization from -50 to -60 mV, whereas the increased sodium pump activity did not appear to influence the acrosome reaction or the mice's overall fertility. A higher $\alpha4$ level is thus sufficient to affect sperm motility and membrane potential, even with the EGFP tag at the C-terminal part of the pump, a positioning that might be expected to cause low sodium affinity for the exogenously expressed pumps [23,24].

Indirect role of $\alpha4$ in maintenance of pH

Intracellular alkalization is one of the prerequisites for sperm capacitation, and the dependence on $\alpha4$ for spermatozoan motility may be due to its indirect role in pH regulation. The sodium gradient drives the transporter family of NHEs (Na^+/H^+-exchangers) to extrude protons in exchange for sodium, which helps to regulate the intracellular pH. Three forms of NHEs have been identified in spermatozoa: NHE1, NHE5 and sperm NHE [25]. Interestingly, male mice lacking the sperm-specific NHE are completely infertile and have severely diminished sperm motility, but the spermatozoa can be rescued by the addition of ammonium chloride (to elevate intracellular pH) and cAMP analogues.

A high mitochondrial activity is necessary in spermatozoa to generate the ATP that fuels flagella motility, giving a constant pressure on the intracellular pH. Incubation with low concentrations of ouabain to specifically inhibit $\alpha4$ reduced the percentage of motile sperm cells from 50 % to ~5 %. The K^+ ionophore valinomycin had no effect, whereas motility was almost completely restored by the drugs nigericin and monensin. The latter ionophores make the membrane permeable to H^+/K^+ and H^+/Na^+ respectively, and would thus

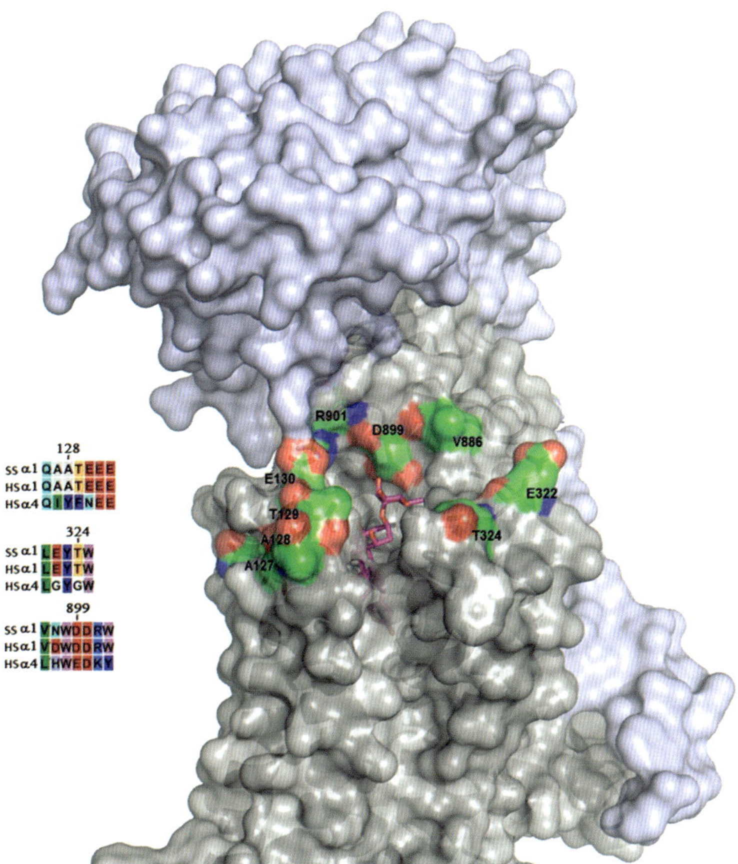

Figure 4. Unique α4 residues near the ouabain-binding pocket
Model of the α1 pump from porcine kidney with ouabain bound (PDB code 3N23). Non-conserved residues around the ouabain-binding pocket are highlighted. The α-subunit is in grey and the β-subunit is in light blue. Ouabain is in magenta. The alignments are *Sus scrofa* (SSα1) and *Homo sapiens* (HSα1 and HSα4).

relieve a possible pH pressure caused by ouabain treatment [26]. It was later confirmed that ouabain inhibition of α4 leads to intracellular acidification [27].

In human sperm, the intracellular pH is further regulated by the proton-selective channel Hv1. Recent advances in electrophysiological measurements directly on spermatozoa have enabled characterization of the sperm Hv1, showing that the channel is activated by membrane depolarization and relative intracellular acidification [28]. Hv1 is not expressed in mouse sperm cells, so genetic studies of the relative roles of NHEs and Hv1 are not applicable to human cells. If Hv1 is the main pH regulator in humans, the mechanism might be expected to depend on the membrane potential rather than on the sodium gradient created by the sodium pump.

Perspectives

For a spermatozoon to fulfil its purpose, a long journey to the egg awaits. To cope with the dramatic environmental alterations it faces, it is equipped with unique membrane proteins that assist in the adjustment to changes in ionic concentrations. In addition, sperm cells need significant amounts of energy to fuel their flagella, but the generation of energy by mitochondria is associated with an acidification of the cytosol that also inhibits the motility of the spermatozoon [25,26]. The sperm-specific sodium pump isoform $\alpha 4$ is therefore believed to be indirectly involved in pH maintenance of mature spermatozoa, and the pump is essential for fertility as shown by both pharmacological inhibition and the recently characterized $\alpha 4$-null mouse. Since $\alpha 4$ is expressed only in spermatozoa, it is an attractive drug target for the development of male contraceptives. The relatively large difference in the affinity towards ouabain for human $\alpha 1$ and $\alpha 4$ suggests an opportunity for the design of ouabain derivatives with high isoform specificity, which may be guided by the crystal structures of ouabain-bound complexes in a potassium-bound low-affinity state [29] and the high-affinity phosphoenzyme state [30]. The β-subunit covers much of the extracellular area of the α-subunit, leaving only a few residues as obvious target motifs (Figure 4). Interestingly, a distinctive $\alpha 4$ motif is readily available for extracellular contact and could be an interesting target site for rational drug development.

Acknowledgements

We thank Natasha Fedosova and Mette Laursen for valuable discussions.

Funding

M.J.C. acknowledges Aarhus Graduate School of Science (AGSoS) and Aarhus Universitets Forskningsfond (AAUF) for financial support.

References

1. Geering, K., Kraehenbuhl, J.P. & Rossier, B.C. (1987) Maturation of the catalytic α-subunit of Na,K-ATPase during intracellular transport. *J. Cell Biol.* **105**, 2613–2619
2. Geering, K. (1991) The functional role of the β-subunit in the maturation and intracellular transport of Na,K-ATPase. *FEBS Lett.* **285**, 189–193
3. Shull, M.M. & Lingrel, J.B. (1987) Multiple genes encode the human Na^+,K^+-ATPase catalytic subunit. *Proc. Natl. Acad. Sci. U.S.A.* **84**, 4039–4043
4. Shamraj, O.I. & Lingrel, J.B. (1994) A putative fourth Na^+,K^+-ATPase α-subunit gene is expressed in testis. *Proc. Natl. Acad. Sci. U.S.A.* **91**, 12952–12956
5. Hlivko, J.T., Chakraborty, S., Hlivko, T.J., Sengupta, A. & James, P.F. (2006) The human Na,K-ATPase $\alpha 4$ isoform is a ouabain-sensitive α isoform that is expressed in sperm. *Mol. Reprod. Dev.* **73**, 101–115
6. Newton, L.D., Kastelic, J.P., Wong, B., van der Hoorn, F. & Thundathil, J. (2009) Elevated testicular temperature modulates expression patterns of sperm proteins in Holstein bulls. *Mol. Reprod. Dev.* **76**, 109–118

7. Woo, A. L., James, P.F. & Lingrel, J.B. (1999) Characterization of the fourth α isoform of the Na,K-ATPase. *J. Membr. Biol.* **169**, 39–44
8. Keryanov, S. & Gardner, K.L. (2002) Physical mapping and characterization of the human Na,K-ATPase isoform, *ATP1A4*. *Gene* **292**, 151–166
9. Underhill, D.A., Canfield, V.A., Dahl, J.P., Gros, P. & Levenson, R. (1999) The Na,K-ATPase α4 gene (*Atp1a4*) encodes a ouabain-resistant α subunit and is tightly linked to the α2 gene (*Atp1a2*) on mouse chromosome 1. *Biochemistry* **38**, 14746–14751
10. Rodova, M., Nguyen, A.N. & Blanco, G. (2006) The transcription factor CREMτ and cAMP regulate promoter activity of the Na,K-ATPase α4 isoform. *Mol. Reprod. Dev.* **73**, 1435–1447
11. Don, J. & Stelzer, G. (2002) The expanding family of CREB/CREM transcription factors that are involved with spermatogenesis. *Mol. Cell. Endocrinol.* **187**, 115–124
12. Wagoner, K., Sanchez, G., Nguyen, A.N., Enders, G.C. & Blanco, G. (2005) Different expression and activity of the α1 and α4 isoforms of the Na,K-ATPase during rat male germ cell ontogeny. *Reproduction* **130**, 627–641
13. Hicks, J.J., Martinez-Manautou, J., Pedron, N. & Rosado, A. (1972) Metabolic changes in human spermatozoa related to capacitation. *Fertil. Steril.* **23**, 172–179
14. Yanagimachi, R. & Bhattacharyya, A. (1988) Acrosome-reacted guinea pig spermatozoa become fusion competent in the presence of extracellular potassium ions. *J. Exp. Zool.* **248**, 354–360
15. Esposito, G., Jaiswal, B.S., Xie, F., Krajnc-Franken, M.A., Robben, T.J., Strik, A.M., Kuil, C., Philipsen, R.L., van Duin, M., Conti, M. et al. (2004) Mice deficient for soluble adenylyl cyclase are infertile because of a severe sperm-motility defect. *Proc. Natl. Acad. Sci. U.S.A.* **101**, 2993–2998
16. Hamamah, S. & Gatti, J.L. (1998) Role of the ionic environment and internal pH on sperm activity. *Hum. Reprod.* **13** (Suppl. 4), 20–30
17. Acott, T.S. & Carr, D.W. (1984) Inhibition of bovine spermatozoa by caudal epididymal fluid. II. Interaction of pH and a quiescence factor. *Biol. Reprod.* **30**, 926–935
18. Blanco, G., Melton, R.J., Sanchez, G. & Mercer, R.W. (1999) Functional characterization of a testes-specific α-subunit isoform of the sodium/potassium adenosinetriphosphatase. *Biochemistry* **38**, 13661–13669
19. Woo, A.L., James, P.F. & Lingrel, J.B. (2000) Sperm motility is dependent on a unique isoform of the Na,K-ATPase. *J. Biol. Chem.* **275**, 20693–20699
20. Sanchez, G., Nguyen, A.N., Timmerberg, B., Tash, J.S. & Blanco, G. (2006) The Na,K-ATPase α4 isoform from humans has distinct enzymatic properties and is important for sperm motility. *Mol. Hum. Reprod.* **12**, 565–576
21. Jimenez, T., Sanchez, G., McDermott, J.P., Nguyen, A.N., Kumar, T.R. & Blanco, G. (2011) Increased expression of the Na,K-ATPase α4 isoform enhances sperm motility in transgenic mice. *Biol. Reprod.* **84**, 153–161
22. Zambrowicz, B.P., Harendza, C.J., Zimmermann, J.W., Brinster, R.L. & Palmiter, R.D. (1993) Analysis of the mouse protamine 1 promoter in transgenic mice. *Proc. Natl. Acad. Sci. U.S.A.* **90**, 5071–5075
23. Morth, J.P., Pedersen, B.P., Toustrup-Jensen, M.S., Sorensen, T.L., Petersen, J., Andersen, J.P., Vilsen, B. & Nissen, P. (2007) Crystal structure of the sodium–potassium pump. *Nature* **450**, 1043–1049
24. Poulsen, H., Khandelia, H., Morth, J.P., Bublitz, M., Mouritsen, O.G., Egebjerg, J. & Nissen, P. (2010) Neurological disease mutations compromise a C-terminal ion pathway in the Na^+/K^+-ATPase. *Nature* **467**, 99–102
25. Wang, D., King, S.M., Quill, T.A., Doolittle, L.K. & Garbers, D.L. (2003) A new sperm-specific Na^+/H^+ exchanger required for sperm motility and fertility. *Nat. Cell Biol.* **5**, 1117–1122
26. Woo, A.L., James, P.F. & Lingrel, J.B. (2002) Roles of the Na,K-ATPase α4 isoform and the Na^+/H^+ exchanger in sperm motility. *Mol. Reprod. Dev.* **62**, 348–356
27. Jimenez, T., Sanchez, G., Wertheimer, E. & Blanco, G. (2010) Activity of the Na,K-ATPase α4 isoform is important for membrane potential, intracellular Ca^{2+}, and pH to maintain motility in rat spermatozoa. *Reproduction* **139**, 835–845
28. Lishko, P.V., Botchkina, I.L., Fedorenko, A. & Kirichok, Y. (2010) Acid extrusion from human spermatozoa is mediated by flagellar voltage-gated proton channel. *Cell* **140**, 327–337

29. Ogawa, H., Shinoda, T., Cornelius, F. & Toyoshima, C. (2009) Crystal structure of the sodium–potassium pump (Na^+,K^+-ATPase) with bound potassium and ouabain. *Proc. Natl. Acad. Sci. U.S.A.* **106**, 13742–13747
30. Yatime, L., Laursen, M., Morth, J.P., Esmann, M., Nissen, P. & Fedosova, N.U. (2010) Structural insights into the high affinity binding of cardiotonic steroids to the Na^+,K^+-ATPase. *J. Struct. Biol.*, doi:10.1016/j.jsb.2010.12.004
31. Levine, N. & Marsh, D.J. (1971) Micropuncture studies of the electrochemical aspects of fluid and electrolyte transport in individual seminiferous tubules, the epididymis and the vas deferens in rats. *J. Physiol.* **213**, 557–570
32. Clemetson, C.A., Kim, J.K., Mallikarjuneswara, V.R. & Wilds, J.H. (1972) The sodium and potassium concentrations in the uterine fluid of the rat at the time of implantation. *J. Endocrinol.* **54**, 417–423
33. Maas, D.H., Stein, B. & Metzger, H. (1984) PO_2 and pH measurements within the rabbit oviduct following tubal microsurgery: reanastomosis of previously dissected tubes. *Adv. Exp. Med. Biol.* **169**, 561–570

Biochem. Soc. Symp. 78
Citation reference: Biochem. Soc. Trans. (2011) **39**, 747–750.

5

Membrane proteins: from bench to bits

Gunnar von Heijne[1]

Center for Biomembrane Research, Department of Biochemistry and Biophysics, Stockholm University, SE-106 91 Stockholm, Sweden, and Science for Life Laboratory Stockholm University, Box 1031, SE-171 21 Solna, Sweden

Abstract

Membrane proteins currently receive a lot of attention, in large part thanks to a steady stream of high-resolution X-ray structures. Although the first few structures showed proteins composed of tightly packed bundles of very hydrophobic more or less straight transmembrane α-helices, we now know that helix-bundle membrane proteins can be both highly flexible and contain transmembrane segments that are neither very hydrophobic nor necessarily helical throughout their lengths. This raises questions regarding how membrane proteins are inserted into the membrane and fold *in vivo*, and also complicates life for bioinformaticians trying to predict membrane protein topology and structure.

Introduction

Until recently, development of methods in the area of membrane protein topology prediction and structure modelling has been based mostly on extracting information from sequence and structure data, but has largely ignored knowledge about how membrane proteins become integrated into membranes and fold *in vivo*. This state of affairs is slowly changing, as we

[1]*email gunnar@dbb.su.se*

now have quantitative data on how much each of the 20 natural amino acids contributes to the apparent free energy of membrane insertion (ΔG_{app}) of isolated transmembrane helices in different biological systems. Such 'biological' hydrophobicity scales have already proved their worth as a basis for topology prediction, and have highlighted that many transmembrane helices in multi-spanning membrane proteins are not sufficiently hydrophobic to be able to insert into the membrane by themselves. Instead, Nature has devised various ways in which other transmembrane helices in the protein can help to ensure that such marginally hydrophobic transmembrane helices end up spanning the membrane.

In the present short review, I first discuss recent approaches to derive experimentally based biological hydrophobicity scales, and then briefly describe how they can be used as a basis for topology prediction. In the final section, I address the problem of marginally hydrophobic transmembrane helices.

Biological hydrophobicity scales

The world is not lacking in hydrophobic scales: a study in 2001 lists about 100 such scales [1] and a simple Google search using the query 'hydrophobicity scale' yields nearly 20000 hits. These scales have been developed with many different uses in mind, but they are all derived either from physical chemistry measurements or from statistical analysis of protein structures or sequences. For those of us interested in membrane proteins, this is not completely satisfactory. Membrane proteins do not partition spontaneously into a pure lipid bilayer *in vivo*, but rather become integrated into a complex biological membrane, in most cases with the help of translocons that provide a very specific environment for the integration step. It is therefore unclear to what extent scales based on, e.g., partitioning of small peptides between aqueous buffer and a liposome or between buffer and an air/water interface are relevant for the *in vivo* assembly of membrane proteins.

Nevertheless, by combining techniques borrowed from molecular cell biology with the systematic approaches typical of physical chemistry, it is possible to derive 'biological' hydrophobicity scales based on measurements of membrane insertion carried out *in vivo* or under conditions that closely mimic the *in vivo* situation. The most complete biological scale to date is that of Hessa et al. [2,3]. The main idea behind this scale is to measure the efficiency of insertion into the mammalian ER (endoplasmic reticulum) membrane of a large panel of model hydrophobic segments placed in the middle of a host membrane protein (Figure 1). Two acceptor sites for N-linked glycosylation bracketing the hydrophobic segment make it easy to determine the degree of membrane insertion, which in turn can be converted into an apparent free energy of membrane insertion (ΔG_{app}). The qualifier 'apparent' is introduced to make clear that it is hard to establish whether the insertion process really represents a global thermodynamic equilibrium.

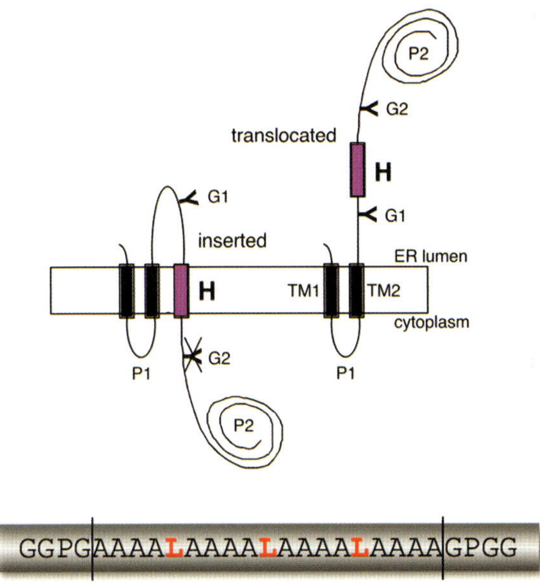

Figure 1. ΔG_{app} values for natural transmembrane helices
The host protein used to measure the efficiency of insertion of a model hydrophobic segment (H) into the ER membrane [2]. The H-segment is flanked by two acceptor sites for asparagine-linked glycosylation (G1 and G2). If the H-segment is inserted as a transmembrane helix, only the G1 site can be modified by the oligosaccharyltransferase in the ER lumen (left). If the H-segment is translocated, both sites are glycosylated. The efficiency of membrane integration can be quantified from SDS/PAGE gels by measuring the amount of singly ($f1$) and doubly ($f2$) glycosylated protein. The apparent free energy of insertion is calculated as $\Delta G_{app} = -RT \ln \frac{f1}{f2}$. A typical H-segment is shown at the bottom.

By scanning each of the 20 naturally occurring amino acids along the model hydrophobic segment, Hessa et al. [2,3] could derive position-specific contributions to ΔG_{app} for all amino acids and could show that these values can be used to rather accurately predict ΔG_{app} for natural transmembrane helices inserted into the host protein.

Not surprisingly, the Hessa scale correlates well with many of the classical hydrophobicity scales, but it has some unexpected qualities. The most puzzling is perhaps that the scale is 'compressed' compared with many other scales in that polar and charged residues increase the free energy of membrane insertion less than expected from simple physical chemistry (i.e. their ΔG_{app} values are less positive than expected) and the non-polar residues decrease the free energy less than expected (i.e. their ΔG_{app} values are less negative than expected). The current working hypothesis is that this is because, first, the high protein content of the ER membrane makes it more polar than a pure lipid bilayer [4], and, secondly, the interior of the Sec61 translocon channel through which the nascent polypeptide is threaded across the membrane is less polar than aqueous buffer [5].

Using similar approaches, biological hydrophobicity scales have also been derived for insertion of transmembrane segments into the ER of the yeast *Saccharomyces cerevisiae* [6] and the inner membrane of *Escherichia coli* [7]; they are both qualitatively similar to the original Hessa scale.

Topology prediction

The development of topology-prediction schemes has followed closely in the footsteps of the hydrophobicity scales. An important advance over the first hydrophobicity-only topology-prediction methods, was the incorporation of the positive-inside rule into the prediction algorithms [8]. Another important step was the use of machine-learning techniques such as hidden Markov models [9,10] and neural networks [11,12].

With machine learning, the number of parameters to be optimized during training of the predictor increase dramatically. Given that the biological hydrophobicity scales offer a possibility to bypass much of the statistics that underlie the commonly used topology-prediction methods, it is interesting to see how well methods based on these scales fare compared with the sophisticated machine-learning predictors. To this end, the TopPred predictor [8], a conceptually very simple algorithm that is based on only a hydrophobicity scale and a simple implementation of the positive-inside rule, was adapted to use the Hessa scale and the algorithm's two adjustable parameters were re-optimized over a training set of proteins with known topologies. The results were encouraging: the new version of TopPred performed more or less on a par with the best machine-learning algorithms [13]. Topology-prediction methods that are based only on experimentally measured parameters thus seem to be within reach, especially if studies of the kind described in the next section will help us better understand how marginally hydrophobic transmembrane helices, which are, by definition, hard to pick up by hydrophobicity analysis, are handled during membrane insertion.

Membrane insertion of marginally hydrophobic helices

The Hessa scale makes it possible to calculate the ΔG_{app} value of any amino acid segment and hence to estimate the membrane-insertion efficiencies of all transmembrane helices in proteins of known structure or topology. As seen in Figure 2, transmembrane helices from single-spanning mammalian membrane proteins all have $\Delta G_{app}<0$, consistent with the fact that they are all efficiently integrated into the ER membrane during synthesis. In contrast, the most hydrophobic segments found in secreted mammalian proteins, i.e. proteins that have all passed through the Sec61 translocon in the ER without being integrated into the membrane, all have $\Delta G_{app}>0$. The ΔG_{app} value thus seems to be a good predictor of membrane insertion propensity. However, a surprisingly large number of transmembrane helices in multi-spanning membrane proteins have

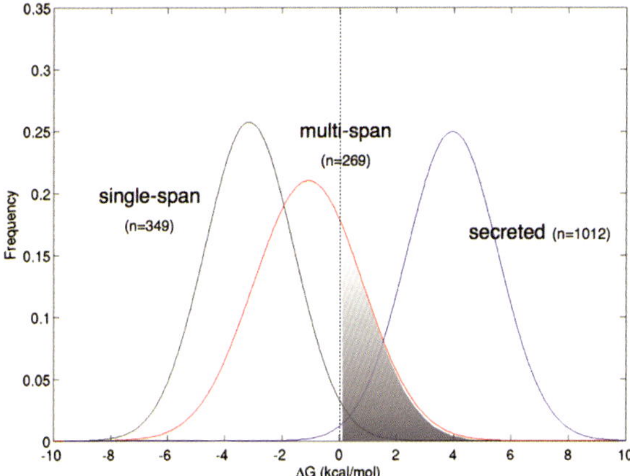

Figure 2. ΔG_{app} value distribution

Distributions of ΔG_{app} values for 349 transmembrane helices from mammalian single-spanning membrane proteins (black curve), for 269 transmembrane helices from multi-spanning membrane proteins of known structure (red curve), and for the most hydrophobic segment found in each of 1012 mammalian secreted proteins (blue curve; signal peptides removed) [3]. The shaded area indicates transmembrane helices in multi-spanning proteins predicted not to insert into the ER membrane. 1 kcal = 4.184 kJ.

$\Delta G_{app}>0$ (shaded grey in Figure 2), i.e. they are predicted not to be able to insert into the ER membrane.

How can this be? The trivial explanation that the predicted ΔG_{app} values are grossly inaccurate can be ruled out: when transmembrane segments from the shaded part of the distribution are tested in the host protein described above, they indeed do not insert into the ER membrane [14]. So far, we have established four ways that such marginally hydrophobic helices can be ‘tricked’ into becoming transmembrane.

The first trick is by adding charged residues to the regions flanking the marginally hydrophobic segment. Positively charged residues have a particularly strong effect, especially when placed near the cytoplasmic end of such a segment (as expected from the positive-inside rule). In a recent study, there was a strong reduction in ΔG_{app} for five out of 15 marginally hydrophobic transmembrane helices chosen from proteins of known structure when the immediate upstream and downstream flanking regions were included [14]. Contributions from flanking residues can therefore explain part of the conundrum.

A second trick was uncovered in an in-depth study of the glutamate-transporter homologue $\mathrm{Glt_{Ph}}$. This protein is a homotrimer and each monomer has eight transmembrane helices, three of which are marginally hydrophobic (Figure 3). TMH4 (transmembrane helix 4) is a particularly odd case: not only is it of low hydrophobicity, but also it has a strange conformation with two breaks in the middle of the membrane. The key to how TMH4 is inserted into the membrane is again found in a flanking sequence, but

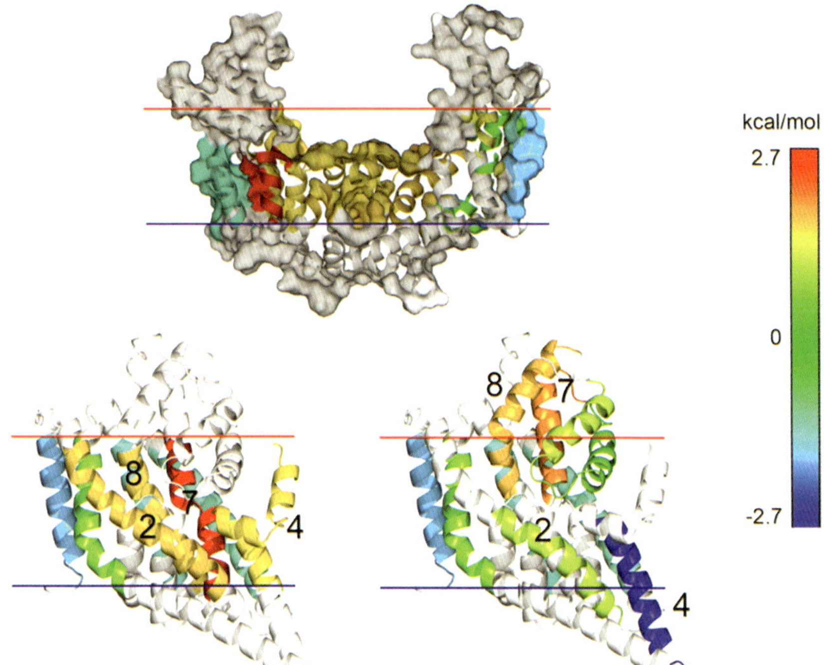

Figure 3 Membrane insertion of transmembrane helices in Glt_{Ph}
A slice through the middle of the Glt_{Ph} trimer (top, PDB code 2NWL) and views of the monomer with transmembrane helices coloured according to ΔG_{app} values (bottom left; see colour scale) and according to ΔG_{app} values of the most hydrophobic stretch overlapping with each of the transmembrane helices (bottom right). Transmembrane helices 2, 4, 7 and 8 are labelled. The very hydrophobic segment (blue) overlapping TMH4 (bottom right) spans the membrane immediately after membrane insertion; presumably, the repositioning of TMH4 relative to the membrane happens during the oligomerization step [15]. 1 kcal = 4.184 kJ.

this time the cytoplasmic flank is strongly hydrophobic instead of charged. Using an approach based on the use of engineered glycosylation sites, it was possible to show that the very hydrophobic segment is the part that is initially inserted into the membrane, implying that the marginally hydrophobic segment is pulled into the membrane only later, presumably during the trimerization step [15]. Repositioning of transmembrane helices relative to the membrane during folding and oligomerization is a second way that marginally hydrophobic transmembrane helices can be formed.

Two additional tricks have been identified, but not studied in detail so far. The first is when charged or polar residues in a neighbouring transmembrane helix form salt bridges or strong hydrogen bonds to an offending polar or charged residue in the marginally hydrophobic helix, thereby locking it in place [16]. The second is when the marginally hydrophobic segment is flanked by two transmembrane helices, both with the same topological preference (for example, both strongly prefer the N_{out}–C_{in} orientation in the membrane). At least in one case, this situation has been shown to force the central, marginally hydrophobic, segment to cross the membrane, in order that both flanking helices can maintain their topologically preferred orientations [17].

Conclusions

Membrane proteins is one area where theoretical and experimental work drive each other in a very productive cycle. Statistical analysis has uncovered regularities such as the positive-inside rule that have in turn inspired experimental work, eventually leading to a deeper understanding of how membrane proteins are assembled *in vivo*, to new approaches to topology prediction, and to insights that will possibly help structure prediction. 'Biological' hydrophobicity scales based on *in vivo* measurements have put the classical physical chemistry of protein–lipid interactions in a new light and have spawned theoretical work trying to better represent the complexity of biological membranes. The rapid progress in the area of membrane protein structural biology seen over the last few years has rejuvenated the whole field and has pinpointed important issues regarding, e.g., how membrane proteins fold and oligomerize that have so far been almost universally ignored. We still have a long way to go before we can confidently say that we understand these greasy little beasts sufficiently well to put them to rest with a contented smile on our face ☺.

Funding

This work was supported by the European Research Council [grant number ERC-2008-AdG 232648], the Swedish Foundation for Strategic Research, the Swedish Research Council, and the Swedish Cancer Foundation.

References

1. Palliser, C.C. & Parry, D.A. (2001) Quantitative comparison of the ability of hydropathy scales to recognize surface β-strands in proteins. *Proteins* **42**, 243–255
2. Hessa, T., Kim, H., Bihlmaier, K., Lundin, C., Boekel, J., Andersson, H., Nilsson, I., White, S.H. & von Heijne, G. (2005) Recognition of transmembrane helices by the endoplasmic reticulum translocon. *Nature* **433**, 377–381
3. Hessa, T., Meindl-Beinker, N.M., Bernsel, A., Kim, H., Sato, Y., Lerch-Bader, M., Nilsson, I., White, S.H. & von Heijne, G. (2007) Molecular code for transmembrane-helix recognition by the Sec61 translocon. *Nature* **450**, 1026–1030
4. Johansson, A.C. & Lindahl, E. (2009) Protein contents in biological membranes can explain abnormal solvation of charged and polar residues. *Proc. Natl. Acad. Sci. U.S.A.* **106**, 15684–15689
5. Schow, E.V., Freites, J.A., Cheng, P., Bernsel, A., von Heijne, G., White, S.H. & Tobias, D.J. (2011) Arginine insertion in membranes: on the connection between molecular dynamics simulations and translocon-mediated insertion experiments. *Mol. Membr. Biol.* **239**, 35–48
6. Hessa, T., Reithinger, J.H., von Heijne, G. & Kim, H. (2009) Analysis of transmembrane helix integration in the endoplasmic reticulum in *S. cerevisiae*. *J. Mol. Biol.* **386**, 1222–1228

7. Xie, K., Hessa, T., Seppälä, S., Rapp, M., von Heijne, G. & Dalbey, R.E. (2007) Features of transmembrane segments that promote the lateral release from the translocase into the lipid phase. *Biochemistry* **46**, 15153–15161
8. von Heijne, G. (1992) Membrane protein structure prediction: hydrophobicity analysis and the positive-inside rule. *J. Mol. Biol.* **225**, 487–494
9. Krogh, A., Larsson, B., von Heijne, G. & Sonnhammer, E. (2001) Predicting transmembrane protein topology with a hidden Markov model: application to complete genomes. *J. Mol. Biol.* **305**, 567–580
10. Tusnady, G.E. & Simon, I. (2001) The HMMTOP transmembrane topology prediction server. *Bioinformatics* **17**, 849–850
11. Rost, B., Fariselli, P. & Casadio, R. (1996) Topology prediction for helical transmembrane proteins at 86% accuracy. *Protein Sci.* **5**, 1704–1718
12. Jones, D.T. (2007) Improving the accuracy of transmembrane protein topology prediction using evolutionary information. *Bioinformatics* **23**, 538–544
13. Bernsel, A., Viklund, H., Falk, J., Lindahl, E., von Heijne, G. & Elofsson, A. (2008) Prediction of membrane-protein topology from first principles. *Proc. Natl. Acad. Sci. U.S.A.* **105**, 7177–7181
14. Hedin, L.E., Ojemalm, K., Bernsel, A., Hennerdal, A., Illergard, K., Enquist, K., Kauko, A., Cristobal, S., von Heijne, G., Lerch-Bader, M. et al. (2009) Membrane insertion of marginally hydrophobic transmembrane helices depends on sequence context. *J. Mol. Biol.* **396**, 221–229
15. Kauko, A., Hedin, L.E., Thebaud, E., Cristobal, S., Elofsson, A. & von Heijne, G. (2010) Repositioning of transmembrane α-helices during membrane protein folding. *J. Mol. Biol.* **397**, 190–201
16. Zhang, L., Sato, Y., Hessa, T., von Heijne, G., Lee, J.K., Kodama, I., Sakaguchi, M. & Uozumi, N. (2007) Contribution of hydrophobic and electrostatic interactions to the membrane integration of the *Shaker* K^+ channel voltage sensor domain. *Proc. Natl. Acad. Sci. U.S.A.* **104**, 8263–8268
17. Ota, K., Sakaguchi, M., von Heijne, G., Hamasaki, N. & Mihara, K. (1998) Forced transmembrane orientation of hydrophilic polypeptide segments in multispanning membrane proteins. *Mol. Cell* **2**, 495–503

Biochem. Soc. Symp. 78
Citation reference: Biochem. Soc. Trans. (2011) **39**, 751–760.

6

Unravelling the folding and stability of an ABC (ATP-binding cassette) transporter

Natalie DiBartolo and Paula J. Booth[1]
School of Biochemistry, University of Bristol, Bristol BS8 1TD, U.K.

Abstract

Prokaryotic importers from the large family of ABC (ATP-binding cassette) transporters comprise four separate subunits: two membrane-embedded and two cytoplasmic ATP-binding subunits. This modular construction makes them ideal candidates for studies of the intersubunit interactions of membrane protein complexes that contain both hydrophobic and hydrophilic subunits. In the present paper, we focus on the vitamin B_{12} importer of *Escherichia coli*, BtuCD, that contains two transmembrane BtuC subunits and two ATP-binding BtuD subunits. We have studied the factors that induce subunit dissociation and unfolding *in vitro*. The BtuCD complex remains intact in alcohol and mild detergents, but urea or SDS separate the BtuC and BtuD subunits, with 6 M urea causing 80% of BtuD to be removed from BtuCD. ATP is found to stabilize the complex as a result of its binding to the BtuD subunits. In the absence of ATP, low concentrations of urea (0.5–3 M) also induce some unfolding, with approximately 14% reduction in helicity in 3 M urea, whereas, in the presence of ATP, no changes are observed. Disassembly at the BtuD–BtuD dimeric interface

[1]*To whom correspondence should be addressed (email paula.booth@bristol.ac.uk).*

in BtuCD can be achieved with smaller concentrations of urea (0.5–3 M) than that required to cause disassembly at the BtuC–BtuD transmission interface (3–8 M), suggesting a stronger interaction of the latter. The results also suggest that unfolding and disassociation of subunits appear to be coupled processes. Our work provides insights into the subunit interactions of an ABC transporter and lays the foundation for studies of the reassembly of BtuCD.

Introduction

The correct folding and assembly of proteins is vital to cells, and unravelling the mechanisms behind these processes has become an intensive area of research. However, work in this field has predominately focused on water-soluble proteins. As a result, our understanding of the factors that determine how α-helical membrane proteins fold is comparatively poor. This is largely because membrane proteins are deemed harder to work with. For many membrane proteins, a major challenge comes with finding appropriate solubilizing systems in which to satisfy their hydrophobicity once outside their native membrane environment [1]. Our knowledge is confined to only a handful of membrane proteins, most notably bacteriorhodopsin, which was the first membrane protein to be refolded from a denatured state *in vitro* [2,3]. Although work on bacteriorhodopsin and other such small monomeric α-helical proteins has resulted in significant breakthroughs, it is important to extend folding studies to include oligomeric membrane proteins which are prevalent in Nature and include the majority of receptors, channels and transporters found in eukaryotes [4,5]. To date, there is a severe lack of information available on the folding and assembly of oligomeric membrane proteins.

ABC (ATP-binding cassette) transporters encompass a large and ubiquitous family of active transporters that use the energy of ATP hydrolysis to translocate a wide variety of substrates across the lipid bilayer of cell membranes, with substrates including sugars, amino acids, lipids, hormones and vitamins [6,7]. Membrane proteins thus lie at the heart of many essential cellular processes. In humans, mutations in the ABC transporter genes result in several genetic diseases such as cystic fibrosis [8], Dublin–Johnson syndrome [9] and Stargardt's disease [10]. Other ABC transporters contribute to the resistance of cancer cells to chemotherapeutic agents [11] and to multidrug and antibiotic resistance in bacterial cells [12]. A typical ABC transporter comprises four core domains: two hydrophobic TMDs (transmembrane domains) and two cytoplasmic NBDs (nucleotide-binding domains) [6,13]. The TMDs provide a pathway for substrates across the lipid bilayer, whereas the NBDs power this transport reaction by binding and hydrolysing ATP. In prokaryotic ABC importers, these domains are generally encoded by up to four separate polypeptide chains, whereas in export systems, they are encoded by two polypeptide chains in prokaryotes or a single chain in eukaryotes. ABC importers also require an additional SBP (substrate-binding protein) which delivers the transport substrate to the TMDs. The structural organization of ABC importers, composed of assemblies of separate domains, makes them ideal systems for studying the assembly and intersubunit interactions of oligomeric membrane proteins [14–17]. In our work, our model system is the *Escherichia coli* ABC transporter, BtuCD.

The vitamin B_{12} import system in *E. coli* encompasses the membrane-associated complex, BtuCD, and the vitamin B_{12}-binding protein, BtuF. Crystal structures are available for both BtuCD and BtuCD-F (BtuCD with BtuF bound) [18,19]. BtuCD functions as a heterotetramer composed of two copies of the TMDs, BtuC, and two copies of the NBDs, BtuD. BtuCD belongs to the type II class of ABC importers which differ in fold and mechanism to those belonging to the type I class of ABC importers. In general, the TMDs of the type II importers have a total of 20 transmembrane helices with ten in each TMD, whereas those of type I contain 12 helices. BtuCD is a particularly interesting model because the requirements for both a highly hydrophobic membrane domain and a hydrophilic aqueous domain must be satisfied during folding and assembly.

Establishing conditions in which BtuCD can be reversibly folded and assembled *in vitro* would provide us with valuable information on BtuCD folding. Therefore, first, conditions must be found with which to disassemble and unfold BtuCD. In our studies, we have investigated urea-induced dissociation and unfolding of BtuCD solubilized in DDM (*n*-dodecyl β-D-maltopyranoside) detergent micelles. The disassembly of BtuCD into its separate building blocks is performed by attaching BtuCD to a nickel-affinity column via a His_{10} tag attached to BtuC and extracting the BtuD subunits (which do not have a His_{10} tag) with urea, leaving BtuC essentially free of BtuD. The efficiency of disassembly under a variety of conditions is assessed by the ability of urea to remove BtuD and is quantified using SDS/PAGE. These experiments have provided insights into the strength and nature of the interactions at the BtuC–BtuD interface, known as the transmission interface. We have also investigated the urea-induced unfolding of BtuCD by monitoring structural and functional changes by the methods of fluorescence, CD spectroscopy and ATPase activity assays. We show that similar concentrations of urea are required to both unfold and disassemble the complex, and thus these processes seem to be coupled. Furthermore, the resistance of BtuCD to chemical denaturant is dependent on whether or not ATP is bound to the nucleotide-binding BtuD domains.

In the present article, we use the following terminology: disassembly, unfolding and denaturation. Disassembly describes changes in the native quaternary structure such as the dissociation of BtuCD at the BtuC–BtuD transmission interface. Unfolding describes reductions in the native tertiary or secondary structure and we also monitor changes in ATPase activity. Finally, denaturation is used to describe any effect that the denaturants have on the protein and may include disassembly and unfolding, as well as any effects on protein function.

Expression and purification of BtuCD

The expression of BtuCD was carried out in *E. coli* BL1(DE3) cells in a fermenter. The purification of the complex from the cell pellets was essentially performed as described previously [18] and the detergent was exchanged from 0.1% LDAO (*N*,*N*-dimethyldodecylamine-*N*-oxide) to 0.1% DDM on a nickel-affinity column. The final buffer used to purify BtuCD contained 25 mM Tris/HCl (pH 7.5), 500 mM NaCl and 0.1% DDM. BtuCD was >98% pure, as judged by SDS/PAGE, with yields of 6–9 mg of BtuCD per 10–15 g of frozen cell pellets. BtuCD carries a His_{10} tag at the N-terminus of BtuC.

Disassembling BtuCD

BtuCD was added to 50% slurry of Ni-NTA (Ni^{2+}-nitrilotriacetate)–agarose beads, pre-equilibrated with BtuCD purification buffer supplemented with 20 mM imidazole, and the mixture was incubated for 1 h at room temperature. The beads were pelleted by centrifugation and the supernatant, containing unbound protein, was removed and retained. The pellet was washed with the above buffer for 5 min at room temperature and re-centrifuged. The pellet was then washed with a series of buffers as above, but containing increasing concentrations of either urea (0.5–2 M, 3–5 M or 6–8 M) or SDS [0.024–0.057% SDS, corresponding to SDS/DDM mole ratios (φ_{SDS}) between 0.3 and 0.5]. The pellet was washed several times in each buffer for 5 min at room temperature, starting with the lowest concentration of denaturant. The beads were pelleted by centrifugation in between each and the supernatants were retained. BtuCD was finally eluted in 500 mM imidazole. In experiments investigating the effects of ATP, disassembly was performed with buffers containing 15 mM ATP and 0.1 mM EDTA or 15 mM ATP and 15 mM $MgCl_2$. The effects of NaCl and glycerol were investigated by carrying out disassembly in 4 M urea and various concentrations of NaCl (0, 0.25, 0.5 and 1 M) or glycerol (10 or 20%, w/v). The presence of BtuD and/or BtuC eluted at different stages of this procedure were analysed by SDS/PAGE using pure BtuCD as a molecular-mass marker for BtuC and BtuD (since BtuCD separates into bands corresponding to BtuC and BtuD on an SDS/polyacrylamide gel). Since only BtuC possesses a His_{10} tag, any BtuD bound to the nickel column must be associated with BtuC.

Densitometric scanning of Coomassie Blue-stained SDS/polyacrylamide gels

Quantification of the amount of BtuD removed was determined by analysis of the Coomassie Blue-stained SDS/polyacrylamide gel using the AlphaErase FC software. Lanes corresponding to pure BtuCD (containing BtuC and BtuD in their original stoichiometric amounts) and the 500 mM imidazole eluent obtained at the end of the disassembly reaction (containing BtuC together with any BtuD unsuccessfully removed) were highlighted for analysis. For each lane, a plot of band intensities against distance (ml) was obtained. In each case, two main peaks were observed corresponding to BtuC and BtuD. The area underneath these peaks was used to calculate the percentage amounts of BtuC and BtuD present before and after disassembly.

Monitoring urea-induced denaturation of BtuCD

CD and fluorescence spectroscopy

BtuCD was diluted in assay buffer containing 25 mM Tris/HCl (pH 7.5), 150 mM NaCl and 0.1% DDM. Various concentrations of urea, dissolved in assay buffer, were added and the mixture was incubated for 10 min at room temperature. In those cases where unfolding was performed in the presence of ATP, 2 mM ATP was added to the mixture before urea. For CD experiments, the final concentration of protein used was 1 mg/ml in a final reaction volume of 50 μl. CD spectra were recorded at Station 12.1 of the SRC (Synchrotron Radiation Source) at Daresbury Laboratories. Spectra were recorded using a 0.1-mm-pathlength cell at 25°C between 200 and 270 nm in 0.5 nm intervals with a 0.5 s integration time and a bandwidth of 1 nm. Data were analysed using the CDtools software [20]. For fluorescence experiments, a final protein concentration of 50 μg/ml was used in a final reaction volume of 200 μl. Fluorescence spectra were collected on a FluoroMax-2 instrument at 25°C using an excitation wavelength of 280 nm and measuring fluorescence emission between 300 and 450 nm using 5 nm emission and excitation bands. Data were fitted using the Microcal Origin software.

ATPase activity assays

Rates of ATP hydrolysis were measured in a final reaction volume of 300 μl containing BtuCD at either 140 or 500 nM and in buffer containing 50 mM Tris/HCl (pH 7.5), 150 mM NaCl and 0.1% DDM. Various concentrations of urea were added, and the mixture was incubated for 10 min at room temperature before transfer to a 37°C waterbath for 3 min. Reactions were initiated by the addition of 2 mM ATP and 10 mM $MgCl_2$. In those experiments where unfolding was investigated in the presence of ATP, ATP was added to the reaction mixture before the urea. In these cases, reactions were initiated by adding 10 mM $MgCl_2$ only. Samples of 50 μl were removed at various time points and added to 50 μl of 12% SDS. ATPase activity was determined by assaying the P_i liberated using the modified molybdate method [21].

Disassembly of BtuCD into BtuC and BtuD with urea and SDS

Upon binding of BtuCD to a nickel-affinity column via the His_{10} tag of BtuC, the column was washed with a series of buffers and their ability to extract and elute BtuD examined. BtuC, together with any BtuD that remained bound to BtuC, was later eluted with imidazole. Disassembly was assessed by SDS/PAGE using purified BtuCD as a molecular-mass marker for separated BtuC and BtuD, since BtuCD separates into three bands on an SDS/polyacrylamide gel: a double band corresponding to BtuC [22] and a single band at a lower molecular mass corresponding to BtuD. The efficiency of dis-assembly was quantified by the percentage amounts of BtuD successfully removed from BtuCD, as determined by densitometric scanning of the SDS/polyacrylamide gels.

Buffers containing glycerol (50%, w/v), methanol (50%, w/v), ethanol (20%, w/v) or OG (octyl β-D-glucoside) (2%, w/v), all at the highest concentrations compatible with the affinity matrix, were unable to induce disassembly and all BtuD remained associated with BtuC (results not shown). Disassembly required harsher buffer conditions containing either urea or SDS, both of which were effective at removing BtuD at certain concentrations. The SDS/polyacrylamide gels shown in Figure 1 follow experiments in which BtuCD was treated with increasing concentrations of either urea or SDS. Although concentrations of urea between 0.5 and 2 M were ineffective at disassembling BtuCD, with only 18% of BtuD eluted overall, significant amounts of BtuD (81%) were eluted with slightly higher urea concentrations of between 3 and 5 M (Figure 1A). Still further increases in the concentration of urea resulted in only marginal increases in the amount of BtuD extracted (88%) (Figure 1A and Table 1). With SDS concentrations of between 0.024 and 0.057% SDS (φ_{SDS} of between 0.3 and 0.5), small amounts of BtuD were removed with 0.039% SDS (φ_{SDS} of 0.4) and, at the end of the experiment, 88% of BtuD was removed (Figure 1B). With both urea and SDS, at the maximal concentrations compatible with the Ni-NTA matrix, small amounts of BtuD remained associated with BtuC and eluted in the final imidazole wash. This residual protein may represent aggregated protein and/or be non-specifically bound to either BtuC or DDM micelles. Alternatively, this may represent a trimeric complex consisting of two BtuC domains and a single BtuD domain, as observed previously by investigations into the gas-phase dissociation pathways of BtuCD in DDM by MS [23]. In this previous study, the BtuD domains were found to dissociate from the BtuCD complex, resulting in a trimeric subcomplex consisting of two BtuC domains and one BtuD domain associated with a DDM detergent micelle.

In the experiments that follow, urea was chosen as the preferred denaturant for disassembling BtuCD. This allows comparisons to be made with other ABC transporters that have previously been disassembled with urea.

The effects of NaCl, glycerol and ATP binding and/or hydrolysis on the disassembly of BtuCD

The nature of the interdomain interactions between BtuC and BtuD was investigated by disassembling BtuCD with urea at various concentrations of either NaCl or glycerol (Table 1). The SDS/polyacrylamide gel in Figure 2(A)

Table 1. Summary of the disassembly of BtuCD under various conditions

For each condition tested, the efficiency of disassembly of BtuCD in DDM is judged by the percentage amount of BtuD successfully removed. Disassembly is scored with either ×, indicating no disassembly, or ✓, indicating disassembly. In the latter case, the more ticks, the greater the efficiency of disassembly. A model of the disassembly is shown in Figure 4(A).

Denaturant	Disassembly conditions	BtuD removed (%)	Efficiency of disassembly
Alcohol	Glycerol (50%, w/v), methanol (50%, w/v), ethanol (20%, w/v)	0	×
OG	2%	0	×
Urea	0.5–2 M	18	✓
	3–5 M	81	✓✓✓
	6–8 M	88	✓✓✓
ATP	3–5 M urea + 15 mM ATP + 0.1 mM EDTA	43	✓✓
	3–5 M urea + 15 mM ATP + 15 mM $MgCl_2$	89	✓✓✓
NaCl	4 M urea + 0 M NaCl	75	✓✓✓
	4 M urea + 0.25 M NaCl	73	✓✓✓
	4 M urea + 0.5 M NaCl	64	✓✓
	4 M urea + 1 M NaCl	30	✓
Glycerol	4 M + 10% glycerol	34	✓
	4 M + 20% glycerol	39	✓
SDS	0.024–0.057%	88	✓✓✓

shows the results from six different experiments, all performed in the presence of 4 M urea, but at various concentrations of NaCl (0, 0.25, 0.5 and 1 M) and glycerol (10 and 20%). For ease of comparison between the experiments, only the eluate from the final imidazole wash is shown (i.e. this contains BtuC with any remaining BtuD, after as much BtuD as possible has been washed away).

BtuCD was found to be stabilized by increasing concentrations of NaCl, as demonstrated by the ability of urea to remove less BtuD and the correspondingly larger amounts eluted in the imidazole wash with BtuC: 75, 73, 64 and 30% of BtuD were removed in the presence of 0, 0.25, 0.5 and 1 M NaCl respectively (Figure 2A). Glycerol also resulted in increased stability, with only half as much BtuD removed as compared with in its absence (both in the presence of 500 mM NaCl): in 10 and 20% glycerol, 34 and 39% of BtuD was removed respectively, whereas, in the absence of glycerol, 64% of BtuD was removed (Figure 2A). Similar results were also observed following disassembly at higher urea concentrations (results not shown).

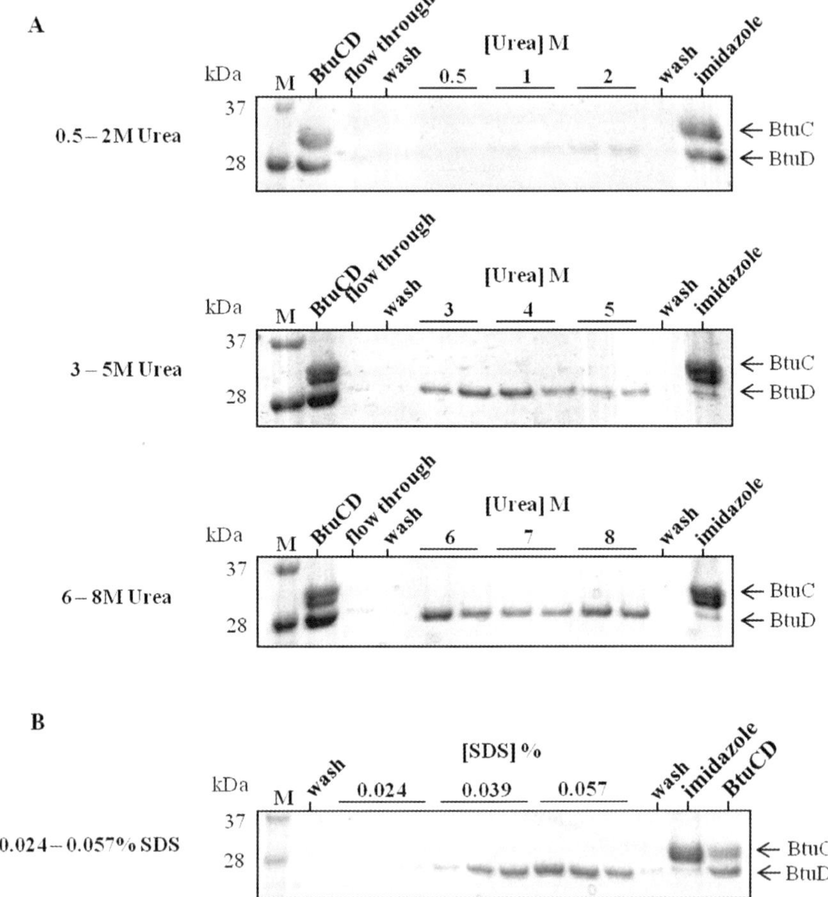

Figure 1. Disassembling BtuCD with urea and SDS

SDS/polyacrylamide gels showing the ability of (**A**) urea and (**B**) SDS to remove BtuD and disassemble BtuCD in DDM. M, molecular-mass standards (sizes are given in kDa); BtuCD, purified BtuCD marker that runs as separate BtuC and BtuD bands on SDS/PAGE; flow through, following binding of BtuCD to the nickel-affinity column; wash, BtuCD buffer wash; imidazole, 0.5 M imidazole wash. Lanes corresponding to washes with increasing concentrations of urea and SDS are labelled accordingly.

Disassembly was also investigated under conditions in which (i) ATP was omitted, (ii) ATP could bind, but not be hydrolysed (ATP/EDTA), and (iii) ATP could bind and be hydrolysed (ATP/Mg^{2+}). In the absence of ATP and in the presence of 15 mM ATP and 15 mM Mg^{2+}, similar amounts of BtuD were eluted from the column with 3–5 M urea: 81 and 89% respectively (Figure 2B). In comparison, the presence of 15 mM ATP and 0.1 mM EDTA stabilized BtuCD, and only half as much BtuD was removed with the same urea concentrations (43%). Similar findings were also observed with concentrations of urea ranging from 6 to 8 M.

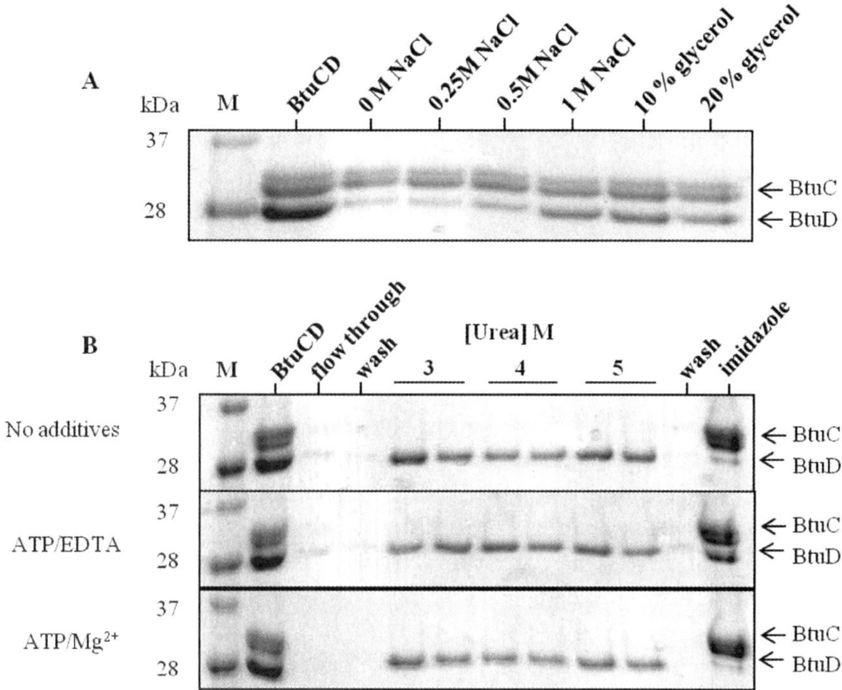

Figure 2. Disassembling BtuCD with urea and SDS
SDS/polyacrylamide gel showing the effects of increasing concentrations of NaCl (0, 0.25, 0.5 and 1 M) and glycerol (10 and 20%, w/v) on the disassembly of BtuCD with 4 M urea. Only samples from the final 0.5 M imidazole eluate are shown from these experiments. (**B**) SDS/polyacrylamide gels following the disassembly of BtuCD with 3–5 M urea in the absence of ATP or in the presence of 15 mM ATP and 0.1 mM EDTA or 15 mM ATP and 15 mM $MgCl_2$. M, molecular-mass standards (sizes are given in kDa); BtuCD, purified BtuCD marker; flow through, following binding of BtuCD to the nickel-affinity column; wash, BtuCD buffer wash; imidazole, 0.5 M imidazole wash. Lanes corresponding to washes with increasing concentrations of urea are labelled accordingly.

Urea induced unfolding of BtuCD

Far-UV CD spectra of BtuCD secondary structure

The far-UV CD spectrum of BtuCD in DDM contains two negative bands at 222 and 208 nm, characteristic of a largely α-helical protein. Owing to the presence of urea in these experiments, data were not collected below a wavelength of 200 nm, as urea absorbs strongly below 210 nm, and qualitative information was obtained from the 222 nm band. Upon treatment with 0.5 M urea, the intensity of the 222 nm band was reduced and further reductions were observed as the concentration of urea was increased, suggesting that urea induces a considerable loss of helical content (Figure 3A and Table 2). In the presence of ATP, the intensity of this band remained constant in up to 3 M urea, suggesting no changes in secondary structure. Under these latter conditions, a reduction in the 222 nm band, and therefore in helical content, was not displayed until urea concentrations of 5 M and higher (Figure 3A). With 7 M urea, similar losses in helical content were observed both with and without ATP, being 33 and 26% respectively.

Fluorescence emission spectra of BtuCD intrinsic tryptophan residues

BtuCD possesses a total of 36 tryptophan residues and displays an intrinsic fluorescence emission spectrum with a maximum at approximately 334 nm in the absence of ATP and 333 nm in the presence of ATP (Figure 3B). In the presence of ATP, the fluorescence signal is 6-fold less than in its absence, suggesting that ATP induces fluorescence quenching. Under both conditions, treatment with 0.5–6 M urea caused a reduction in the fluorescence intensity coupled with a slight red-shift in the position of the emission maximum (Figure 3B). These changes are consistent with tryptophan residues moving from a more to less hydrophobic environment and thus of unfolding. In the absence of ATP, the magnitude of the red-shift was more than twice that observed in its presence, being 2.7 and 1.1 nm respectively (Table 2).

ATPase activity assays of BtuCD

Activity assays were performed at two different protein concentrations (140 and 500 nM) and following incubation with 0–3 M urea (with or without ATP) where the BtuCD complex remains largely intact (see above). At both protein concentrations, higher rates of ATP hydrolysis were displayed following treatment with 0.5–2 M urea in the presence of ATP compared with in the absence of ATP (Figure 3C). For example, incubation of 140 nM BtuCD with 0.5 M urea in the presence of ATP resulted in a rate of ATP hydrolysis of 565 nmol/min per mg, whereas incubation with the same urea concentration, but in the absence of ATP, resulted in a rate of just 108 nmol/min per mg. These rates correspond to a reduction in ATPase activity of 80 and 1% respectively (as compared with the activity of folded BtuCD in the absence of denaturant). Virtually all activity was abolished with 3 M urea, both with and without ATP. With certain urea concentrations (0.5 and 1 M), there is a clear correlation between the concentration of protein used in the assay and the amount of activity remaining, with higher protein concentrations resulting in higher rates of ATP hydrolysis (Figure 3C). For example, incubation of 500 nM BtuCD with 0.5 M urea without ATP resulted in an ATPase activity of 376 nmol/min per mg, approximately three times higher than the activity displayed by 140 nM BtuCD under the same conditions (108 nmol/min per mg). The effects of protein concentration on the rates of ATP hydrolysis following urea treatment in the presence of ATP are less clear.

Taken together, these data suggest that urea is able to induce unfolding of BtuCD. Furthermore, BtuCD is more resistant to urea-induced unfolding when ATP is bound to the nucleotide-binding BtuD domains.

Table 2. Summary of the structural and functional changes observed during the unfolding experiments of BtuCD in urea

A model of the unfolding is shown in Figure 4(B).

	Change in CD signal at 222 nm (%)		Red-shift in fluorescence emission maximum (λ_{max}) (nm)		Reduction in ATPase activity of folded BtuCD (%)			
[Urea] (M)	−ATP	+ATP	−ATP	+ATP	140 nM BtuCD −ATP	500 nM BtuCD −ATP	140 nM BtuCD +ATP	500 nM BtuCD +ATP
0.5	4	1	0.5	0	80	42	1	11
1	9	0	0.6	0.4	95	86	6	27
2	–	–	1.4	0.5	95	99	82	63
3	14	0	1.9	0.4	98	99	100	97
4	–	–	2.3	0.8	–	–	–	–
5	20	6	2.4	0.9	–	–	–	–
6	–	23	2.7	1.1	–	–	–	–
7	26	33	–	–	–	–	–	–

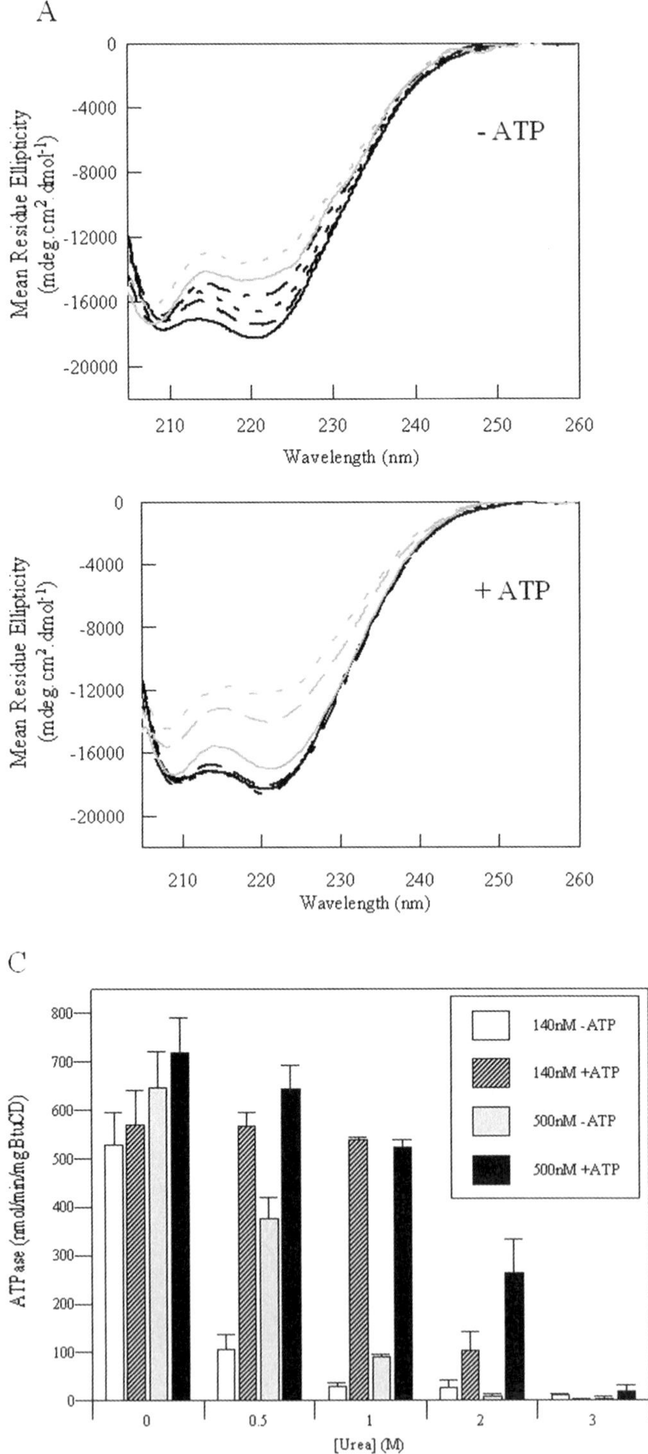

Figure 3. **See page 73 for caption**

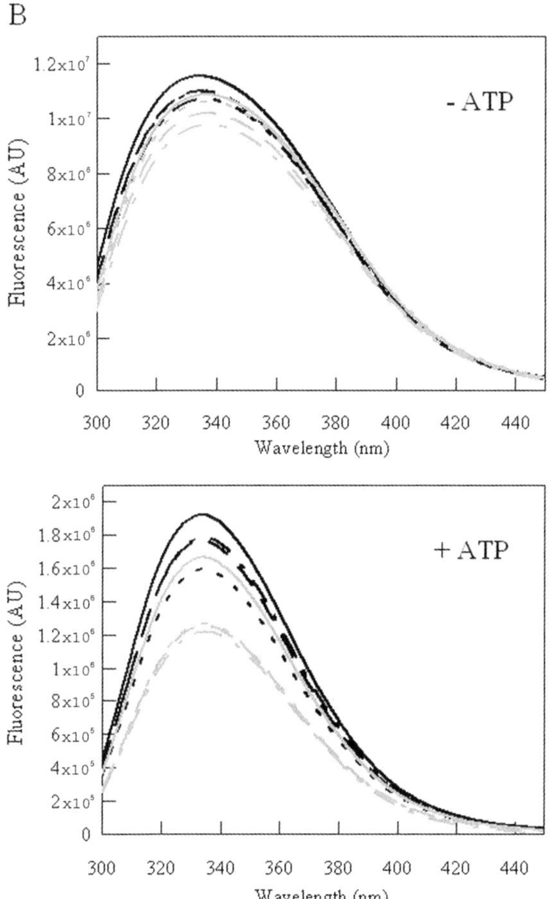

Figure 3. Urea-induced unfolding of BtuCD

(**A**) Far-UV CD spectra of BtuCD in the absence (left) or presence of 2 mM ATP (right) in 0M (continuous black line), 0.5 M (dashed black line), 1 M (dotted black line), 3 M (dashed-dotted black line), 5 M (solid grey line), 6 M (dashed grey line) and 7 M (dotted grey line) urea. (**B**) Fluorescence emission spectra of BtuCD in the absence (left) and presence of 2 mM ATP (right) in 0 M (continuous black line), 0.5 M (dashed black line), 1 M (dotted black line), 2 M (dashed-dotted black line), 3 M (solid grey line), 4 M (dashed grey line), 5 M (dotted grey line), 6 M (dashed-dotted grey line) urea. (**C**) ATPase activity of BtuCD following treatment with urea concentrations of between 0 and 3 M.

The strength and nature of protein–protein interactions at the transmission interface

Disassembly experiments, performed by attaching BtuCD to an Ni-NTA column (via the His_{10} tag of BtuC), have provided valuable insights into the strength and nature of the interactions at the BtuC–BtuD interface, known

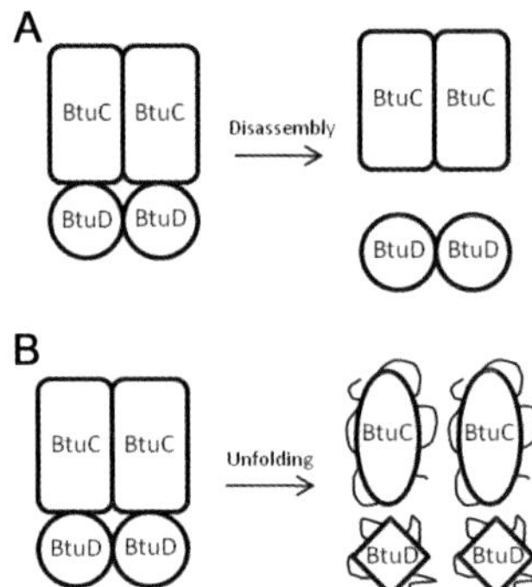

Figure 4. Models of disassembly (A) and unfolding (B) of BtuCD

as the transmission interface. The interface buried between these domains is large, with a total surface area of ~1500 Å^2 (1 Å = 0.1 nm) [18]. A strong protein–protein interaction is observed which cannot be perturbed under 'mild' conditions such as alcohols or a mild detergent, OG. To separate these domains, considerably harsher conditions of SDS and urea are required, both of which are well-known denaturants of water-soluble proteins and have, in some cases, also been used to successfully denature membrane proteins. Treatment of BtuCD with urea (3–8 M) and SDS (0.024–0.057%, corresponding to φ_{SDS} 0.3–0.5) results in the extraction of the majority of BtuD, leaving BtuC largely devoid of BtuD, as determined by the quantification of small amounts of BtuD from SDS/polyacrylamide gels. With 6–8 M urea, 80% of BtuD is removed and these data are consistent with results obtained for two other ABC transporters of *Salmonella enterica* serotype Typhimurium in inside-out membrane vesicles: the maltose ($MalFGK_2$) and histidine ($HisQMP_2$) transporters. For the maltose transporter, 69% of the nucleotide-binding MalK domains were removed with 6.6 M urea [15] and 49% of the homologous HisP domains of the histidine transporter were removed with 7.3 M urea [17]. In each case, small amounts of the NBDs remained associated with their transmembrane counterparts and have been suggested to represent aggregated protein and/or may be non-specifically bound to either the TMDs or the column [15]. Our results also show that the presence of ATP/Mg^{2+} does not alter the ability of urea to remove BtuD, and similar findings have been observed for the maltose transporter [15]. Contrasting results are reported for the histidine transporter, and, under the same conditions, urea is able to extract essentially all of HisP [17]. This led to the conclusion that HisP is tightly engaged with the HisQM TMDs, but becomes disengaged and more solvent-exposed upon binding ATP/Mg^{2+} [15,17,24]. Evidently, this is not the case for BtuCD and may reflect structural and/or mechanistic differences between the ABC transporters. Using conditions under which ATP could bind but not be hydrolysed (ATP/EDTA), approximately

50% less BtuD (43%) was removed with 3–5 M urea than in the absence of ATP (88%) or using conditions under which ATP hydrolysis was allowed to proceed (ATP/Mg^{2+}) (89%).

Hydrophobic interactions appear to have an important stabilizing role at the transmission interface, as judged by the ability of urea to remove less BtuD in increasing salt concentrations, with 64 and 34% of BtuD removed with 4 M urea in 0.5 M and 1 M NaCl respectively. Less BtuD was also extracted with 4 M urea in the presence of glycerol, which weakens hydrophobic interactions, with 64% and 34% of BtuD removed in the absence of glycerol and 10% (w/v) glycerol respectively. However, glycerol is also known to enhance the structure and stability of proteins through a number of other factors, including the preferential hydration of the protein, enhanced internal hydrogen-bonding and co-solvent exclusion. Therefore any one, or a combination, of these factors could account for the increased stability of BtuCD in urea in the presence of glycerol. The significance of hydrophobic interactions at the BtuC–BtuD interface is also supported by the available crystal structure of BtuCD [18]. On searching for all residues at the transmission interface with an atom within 5 Å of a residue in the other domain, two clear hydrophobic patches are revealed with residues positioned in both domains that would result in the formation of hydrophobic interactions. Space-filling models displaying the interacting surfaces of each domain show that, although the surface of BtuC is considerably more hydrophobic, there are distinct and matching regions of hydrophobicity on the surface of BtuD.

Urea-induced denaturation of BtuCD and the stabilizing effects of ATP

The ability of urea to denature native BtuCD to a state of structural and functional deformity has been confirmed using CD and fluorescence spectroscopy as well as ATPase activity assays. The effects of ATP binding on this process have also been examined. In the absence of ATP, low urea concentrations (0.5–3 M) are able to induce some unfolding. In 3 M urea, BtuCD loses approximately 14% of its helical content, as measured by the reduction in the CD signal at 222 nm, and possesses tryptophan residues that are more solvent-exposed. The binding of ATP leads to increased structural stability and, under the same conditions, no changes in CD signal at 222 nm are observed and tryptophan residues are less solvent-exposed. These concentrations of urea also result in a reduction in the ability of BtuCD to hydrolyse ATP and, in 3 M urea, essentially all activity is abolished. Two ATP-binding sites form at the interface between two BtuD domains. When part of the complete BtuCD complex solubilized in DDM, these domains hydrolyse ATP at a basal rate of ~640 nmol/min per mg, corresponding to a turnover of 1.39 s^{-1}. This rate of ATP hydrolysis is for BtuCD in the absence of the vitamin

B_{12}-binding protein, BtuF, as no stimulation of activity was seen for BtuCD in the presence of BtuF. The isolated BtuD domains display a basal rate of ATP hydrolysis of ~193 nmol/min per mg (0.2 s^{-1}), approximately 30% of that displayed when part of the complete complex. The amount of activity remaining following incubation of BtuCD with urea in the absence of ATP is dependent on protein concentration, with higher protein concentrations resulting in greater rates of ATP hydrolysis. As the active form of BtuCD involves the dimerization of two BtuD domains, this result is consistent with urea causing dissociation of the functional BtuD dimer. Greater rates of ATP hydrolysis are also observed following incubation of BtuCD with urea in the presence of ATP as compared with in the absence of ATP. This supports further the notion that ATP serves to stabilize BtuCD in urea. Under these latter conditions (in the presence of ATP), although dissociation of the BtuD dimer is still thought to occur, ATP binding across the dimer interface is likely to provide the dimer with additional stability. These findings are also consistent with the gas-phase dissociation pathways observed by MS of the complete BtuCD complex ($BtuC_2D_2$) in DDM [23]: at low activation energies and in the absence of ATP, BtuD is released from the complex and the trimeric dissociation product, $BtuC_2D$, is formed. However, upon ATP binding, this dissociation product is not observed, which is in line with an increase in interactions between two BtuD domains preventing dissociation and stabilizing the nucleotide-bound form of $BtuC_2D_2$. The BtuD dimer interface consists mainly of residues from the Walker A-motif, D-loop and switch region [18]. In the presence of ATP, additional contacts are made through the binding of two ATP molecules between the Walker A motif of one domain and the ABC signature motif of the other domain. Under these conditions, low urea concentrations are thought to disrupt these interactions at the BtuD–BtuD interface, without causing any unfolding in the individual BtuD domains. Changes in the quaternary structure of BtuCD, such as the dissociation of the BtuD dimer, would not be expected to be detected by CD measurements and indeed no changes were observed for BtuCD in 0.5–3 M urea in the presence of ATP. Other experiments carried out in our laboratory with the isolated BtuD domains, purified as a soluble protein, have confirmed that BtuD displays remarkable resistance in urea when ATP is bound and does not behave as a typical water-soluble protein (results not shown). Changes in the CD spectra of BtuCD are observed, however, following incubation with these low urea concentrations (0.5–3 M) in the absence of ATP, suggesting that the dissociation of the BtuD dimer is accompanied by unfolding of the separated BtuD domains.

Disassembly at the BtuD–BtuD dimeric interface in BtuCD can be achieved with smaller concentrations of urea (0.5–3 M) than that required to cause disassembly at the BtuC–BtuD transmission interface (3–8 M), suggesting that the interdomain interactions between the two BtuD domains are considerably weaker than those between BtuC and BtuD. This is consistent with the crystal structure of BtuCD which shows that the interface buried between the two

BtuD domains (~750 Å^2) is much smaller than that buried between the BtuC and BtuD domains (~1500 Å^2) [18].

Unfolding of BtuCD and dissociation at the BtuC–BtuD interface are coupled processes

In our experiments, the ability of urea to (i) disassemble the BtuCD complex at the transmission interface, and (ii) induce unfolding of the complete BtuCD complex has been investigated. Combining the results from these experiments provides insights into the mechanisms by which BtuCD unfolds and disassembles *in vitro*. One key feature that has emerged is that unfolding and disassembly appear to be coupled: upon binding ATP, BtuCD is stabilized in urea against both unfolding and dissociation at the BtuC–BtuD interface. Conversely, in the absence of ATP, both the extent of unfolding and the efficiency of disassembly are considerably greater.

Coupled unfolding and dissociation has previously been observed for KcsA, the bacterial K^+ channel. KcsA consists of a highly stable heterotetramer which can be unfolded with TFE (2,2,2-trifluoroethanol) [4,25]. At low TFE concentrations (22–30%, v/v) KcsA loses a significant amount of helical structure and tryptophan residues move to a more polar environment. The same TFE concentrations were also found to cause dissociation of the native tetramer into monomers, as determined by SDS/PAGE (as KcsA migrates as a tetramer) and AUC (analytical ultracentrifugation) experiments [4]. Linked folding and tetramerization events have also been reported for the T1 domain of the eukaryotic K^+ channel, *Shaker* B [26]. The authors suggest that, although some tertiary structure folding of individual monomers may occur in the absence of tetramer formation, further tertiary folding is accompanied by tetramerization. This coupling is suggested to serve to stabilize tertiary structure by quaternary structure or vice versa. Coupling between tertiary and quaternary structure may therefore be a common theme for oligomeric protein complexes and indeed this appears true for BtuCD.

Funding

This work was funded by a Biotechnology and Biological Sciences Research Council Ph.D. studentship to N.D.B. and the European Union via the European Membrane Protein Consortium (EMeP). P.J.B. also holds a Royal Society Wolfson Merit Award.

References

1. Booth, P.J., Templer, R., Meijberg, W., Allen, S., Curran, A. & Lorch, M. (2001) *In vitro* studies of membrane protein folding. *Crit. Rev. Biochem. Mol.* **36**, 501–503
2. London, E. & Khorana, H.G. (1982) Denaturation and renaturation of bacteriorhodopsin in detergents and lipid–detergent mixtures. *J. Biol. Chem.* **257**, 7003–7011

3. Booth, P.J. & Curnow, P. (2006) Membrane proteins shape up: understanding *in vitro* folding. *Curr. Opin. Struct. Biol.* **16**, 480–488
4. Barrera, F.N., Renart, M.L., Molina, M.L., Poveda, J.A., Encinar, J.A., Fernandez, A.M., Neira, J.L. & Gonzalez-Ros, J.M. (2005) Unfolding and refolding *in vitro* of a tetrameric, α-helical membrane protein: the prokaryotic potassium channel Kcs. *Biochemistry* **44**, 14344–14352
5. Gorzelle, B.M., Nagy, J.K., Oxenoid, K., Lonzer, W.L., Cafiso, D.S. & Sanders, C.R. (1999) Reconstitutive refolding of diacylglycerol kinase, an integral membrane protein. *Biochemistry* **38**, 16373–16382
6. Higgins, C.F. (1992) ABC transporters: from microorganisms to man. *Annu. Rev. Cell Biol.* **8**, 67–113
7. Biemans-Oldehinkel, E., Doeven, M.K. & Poolman, B. (2006) ABC transporter architecture and regulatory roles of accessory domains. *FEBS Lett.* **580**, 1023–1035
8. Collins, F.S. (1992) Cystic fibrosis: molecular biology and therapeutic implications. *Science* **256**, 774–779
9. Rust, S., Rosier, M., Funke, H., Real, J., Amoura, Z., Piette, J.C., Deleuze, J.F., Brewer, H.B., Duverger, N., Denèfle, P. & Assmann, G. (1999) Tangier disease is caused by mutations in the gene encoding ATP-binding cassette transporter 1. *Nat. Genet.* **22**, 352–355
10. Allikmets, R., Singh, N., Sun, H., Shroyer, N.F., Hutchinson, A., Chidambaram, A., Gerrard, B., Baird, L., Stauffer, D., Peiffer, A. et al. (1997) A photoreceptor cell-specific ATP-binding transporter gene (ABCR) is mutated in recessive Stargardt macular dystrophy. *Nat. Genet.* **15**, 236–246
11. Gottesman, M.M. & Pastan, I. (1993) Biochemistry of multidrug resistance mediated by the multidrug transporter. *Annu. Rev. Biochem.* **62**, 385–427
12. Jones, P.M. & George, A.M. (2004) The ABC transporter structure and mechanism: perspectives on recent research. *Cell. Mol. Life Sci.* **61**, 682–699
13. Locher, K.P. (2009) Structure and mechanism of ATP-binding cassette transporters. *Philos. Trans. R. Soc. London Ser. B* **364**, 239–245
14. Mourez, M., Jéhanno, M., Schneider, E. & Dassa, E. (1998) *In vitro* interaction between components of the inner membrane complex of the maltose ABC transporter of *Escherichia coli*: modulation by ATP. *Mol. Microbiol.* **30**, 353–363
15. Landmesser, H., Stein, A., Blüschke, B., Brinkmann, M., Hunke, S. & Schneider, E. (2002) Large-scale purification, dissociation and functional reassembly of the maltose ATP-binding cassette transporter (MalFGK$_2$) of *Salmonella typhimurium*. *Biochim. Biophys. Acta* **1565**, 64–72
16. Sharma, S., Davis, J.A., Ayvaz, T., Traxler, B. & Davidson, A.L. (2005) Functional reassembly of the *Escherichia coli* maltose transporter following purification of a MalF–MalG subassembly. *J. Bacteriol.* **187**, 2908–2911
17. Liu, P.Q. & Ames, G.F. (1998) *In vitro* disassembly and reassembly of an ABC transporter, the histidine permease. *Proc. Natl. Acad. Sci. U.S.A.* **95**, 3495–34500
18. Locher, K.P., Lee, A.T. & Rees, D.C. (2002) The *E. coli* BtuCD structure: a framework for ABC transporter architecture and mechanism. *Science* **296**, 1091–1098
19. Hvorup, R.N., Goetz, B.A., Niederer, M., Hollenstein, K., Perozo, E. & Locher, K.P. (2007) Asymmetry in the structure of the ABC transporter-binding protein complex BtuCD–BtuF. *Science* **317**, 1387–1390
20. Lees, J.G., Smith, B.R., Wien, F., Miles, A.J. & Wallace, B.A. (2004) CDtool: an integrated software package for circular dichroism spectroscopic data processing, analysis, and archiving. *Anal. Biochem.* **332**, 285–289
21. Chifflet, S., Torriglia, A., Chiesa, R. & Tolosa, S. (1988) A method for the determination of inorganic phosphate in the presence of labile organic phosphate and high concentrations of protein: application to lens ATPases. *Anal. Biochem.* **168**, 1–4
22. Borths, E.L., Locher, K.P., Lee, A.T. & Rees, D.C. (2002) The structure of *Escherichia coli* BtuF and binding to its cognate ATP binding cassette transporter. *Proc. Natl. Acad. Sci. U.S.A.* **99**, 16642–16647
23. Barrera, N.P., Bartolo, N.D., Booth, P.J. & Robinson, C.V. (2008) Micelles protect membrane complexes from solution to vacuum. *Science* **321**, 243–246

24. Liu, P.Q., Liu, C.E. & Ames, G.F. (1999) Modulation of ATPase activity by physical disengagement of the ATP-binding domains of an ABC transporter, the histidine permease. *J. Biol. Chem.* **274**, 18310–18318
25. Barrera, F.N., Renart, M.L., Poveda, J.A., de Kruijff, B., Killian, J.A. & Gonzalez-Ros, J.M. (2008) Protein self-assembly and lipid binding in the folding of the potassium channel KcsA. *Biochemistry* **47**, 2123–2133
26. Robinson, J.M. & Deutsch, C. (2005) Coupled tertiary folding and oligomerization of the T1 domain of K_v channels. *Neuron* **45**, 223–232

Biochem. Soc. Symp. 78
Citation reference: Biochem. Soc. Trans. (2011) **39**, 761–766.

7

Lipid–protein interactions

Anthony G. Lee[1]

School of Biological Sciences, Life Sciences Building 85, University of Southampton, Southampton SO17 1BJ, U.K.

Abstract

Intrinsic membrane proteins are solvated by a shell of lipid molecules interacting with the membrane-penetrating surface of the protein; these lipid molecules are referred to as annular lipids. Lipid molecules are also found bound between transmembrane α-helices; these are referred to as non-annular lipids. Annular lipid binding constants depend on fatty acyl chain length, but the dependence is less than expected from models based on distortion of the lipid bilayer alone. This suggests that hydrophobic matching between a membrane protein and the surrounding lipid bilayer involves some distortion of the transmembrane α-helical bundle found in most membrane proteins, explaining the importance of bilayer thickness for membrane protein function. Annular lipid binding constants also depend on the structure of the polar headgroup region of the lipid, and hotspots for binding anionic lipids have been detected on some membrane proteins; binding of anionic lipid molecules to these hotspots can be functionally important. Binding of anionic lipids to non-annular sites on membrane proteins such as the potassium channel KcsA can also be important for function. It is argued that the packing preferences of the membrane-spanning α-helices in a membrane protein result in a structure that matches nicely with that of the surrounding lipid bilayer, so that lipid and protein can meet without either having to change very much.

[1]*email agl@soton.ac.uk*

A wide variety of lipids

Most biological membranes contain a complex mixture of lipid molecules differing in their polar regions and in their fatty acyl chains, the structures of the lipid molecules being different in membranes from different organisms. Furthermore, the lipid composition of a membrane is not held totally constant, and, for example, in animals varies with diet and in bacteria varies with growth conditions such as temperature. This is unusual in biology; important structures are usually conserved so that, for example, the protein carrying out some particular function in one type of cell is very similar to the protein carrying out the same function in another type of cell. Clearly, the specific chemical structures of the lipid molecules in biological membranes have not been conserved. What is conserved is an amphipathic structure capable of forming bilayers in water, with some conserved features. One such conserved feature, true for most, if not all, membranes, is that the bilayer is fluid, a state achieved if the lipid molecules contain unsaturated fatty acyl chains. A second conserved feature is the hydrophobic thickness of the bilayer (the thickness of the fatty acyl chain region), which, although not identical for all membranes, does not vary much from approximately 25 to 30 Å (1 Å = 0.1 nm). The cost of exposing hydrophobic groups to water is high, so that the hydrophobic thickness of the bilayer component of a membrane is expected to match the hydrophobic thicknesses of the proteins in that membrane.

The importance of hydrophobic matching

The importance of hydrophobic matching can be demonstrated very clearly using reconstituted membrane systems. Figure 1 shows the effect of fatty acyl chain length on the rate of hydrolysis of ATP by the Ca^{2+}-ATPase from skeletal muscle sarcoplasmic reticulum reconstituted into bilayers of phosphatidylcholines in the liquid crystalline phase [1–3]. Highest activity is seen with a fatty acyl chain length of about C_{18}, with lower activities in bilayers of lipids with shorter or longer fatty acyl chains.

The shape of the activity plot shown in Figure 1 can be compared with the chain-length-dependence of the energetic cost of distorting the lipid bilayer around the Ca^{2+}-ATPase to achieve hydrophobic matching. Marsh [4] has shown that the free energy ΔG_{el} required to stretch or compress a lipid molecule around a membrane protein to achieve hydrophobic matching can be expressed as:

$$\Delta G_{el} = 15.1 \frac{(l_L - l_o)^2}{l_o}$$

where l_o is the undistorted hydrophobic length of the lipid and l_L is the distorted thickness required for hydrophobic matching; the units for ΔG_{el} are kJ per lipid molecule and the units of length are nm. Marsh [4] has also shown that

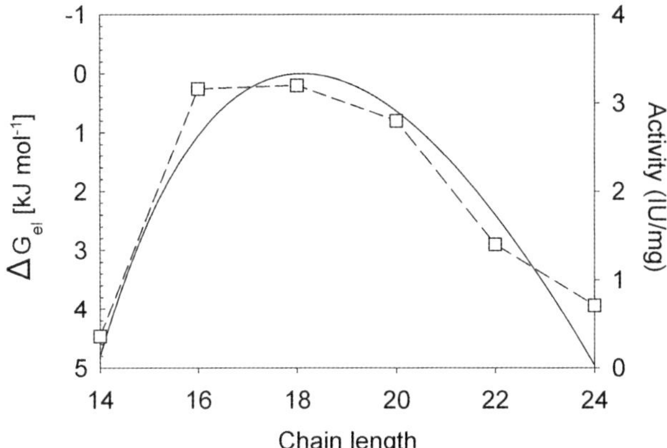

Figure 1. The effect of fatty acyl chain length on the ATPase activity of the Ca^{2+}-ATPase

Ca^{2+}-ATPase was reconstituted into bilayers of mono-unsaturated phosphatidylcholines. (□, right-hand axis) and ATPase activities were measured at 25°C as a function of fatty acyl chain length. Free energies of lipid distortion ΔG_{el} (solid continuous line, left-hand axis) were calculated as described in the text.

the hydrophobic thickness of a bilayer of a phosphatidylcholine with mono-unsaturated chains in the liquid crystalline phase, at 30°C is given by:

$$d_L = 1.9(n_C - 3.9)$$

where d_L is the hydrophobic thickness in Å and n_C is the number of carbons per chain. These two equations can be used to calculate the lipid distortion energy as a function of chain length for a protein of given hydrophobic thickness. As the optimal activity for the Ca^{2+}-ATPase is observed in bilayers of di($C_{18:1}$)PC (dioleoylphosphatidylcholine), it is reasonable to assume that this lipid provides best matching to the protein, giving a hydrophobic thickness for the protein of 27 Å. The distortion energy required to achieve hydrophobic matching will then increase as the bilayer thickness diverges from the optimal value of 27 Å. As shown in Figure 1, the dependence of ΔG_{el} on chain length is remarkably similar to the dependence of activity on chain length.

Effects of hydrophobic mismatch are complex

The obvious conclusion from Figure 1 is that there is a direct relationship between ATPase activity and lipid distortion energy, but this would be incorrect; the relationship is more complex than first appears. In particular, the reasons for activities being low in short chain lipids are different from the reasons for their being low in long-chain lipids (Figure 2). The low rate of ATP hydrolysis in di($C_{14:1}$)PC (dimyristoleoylphosphatidylcholine) follows in large part from a low rate of phosphorylation of the Ca^{2+}-ATPase by ATP [5]. In contrast, in

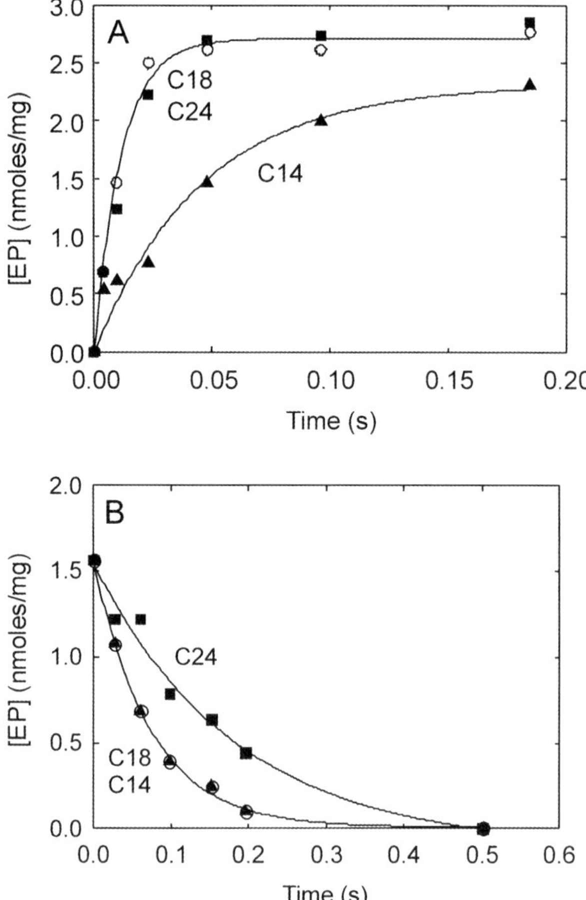

Figure 2. Effects of fatty acyl chain length on the rates of phosphorylation and dephosphorylation of the Ca^{2+}-ATPase by ATP

The ATPase was reconstituted into bilayers of di($C_{14:1}$)PC (▲), di($C_{18:1}$)PC (○) or di($C_{24:1}$)PC (■). (**A**) Rate of phosphorylation. The reconstituted ATPase was mixed with 50 μM ATP, and the level of phosphoenzyme (EP) was determined as a function of time at 25°C. (**B**) Rate of dephosphorylation. The reconstituted ATPase was first phosphorylated with phosphate at pH 6.0 and then mixed with an excess of pH 7.5 buffer containing ATP to induce dephosphorylation, and the level of phosphoenzyme (EP) was determined as a function of time at 25°C.

di($C_{24:1}$)PC (dinervonylphosphatidylcholine), the rate of phosphorylation is the same as that in di($C_{18:1}$)PC, and the low rate of ATP hydrolysis by the Ca^{2+}-ATPase in di($C_{24:1}$)PC follows from a low rate of dephosphorylation of the phosphorylated intermediate [5,6] (Figure 2).

The simplest explanation for the observed effects of lipid chain length on Ca^{2+}-ATPase function would be that the reaction mechanism for the Ca^{2+}-ATPase remains the same, but that changing lipid chain length results in a

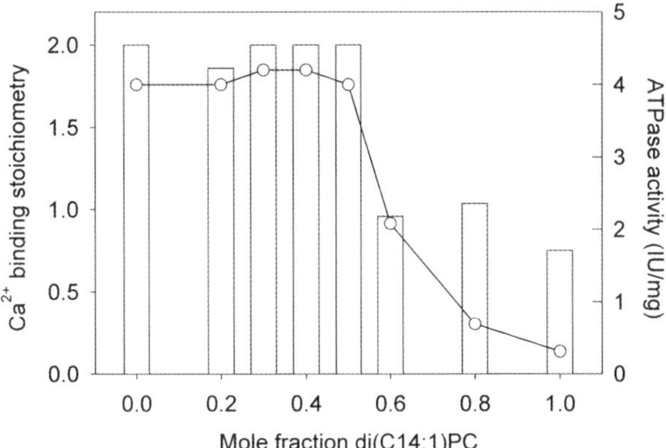

Figure 3. Effects of mixtures of two phospholipids of different chain lengths on Ca^{2+} binding and activity of the Ca^{2+}-ATPase

The Ca^{2+}-ATPase was reconstituted with mixtures of di($C_{14:1}$)PC and di($C_{18:1}$)PC containing the indicated mole fraction of di($C_{14:1}$)PC. The bars show the stoichiometry of Ca^{2+} binding to the ATPase. Also shown are the ATPase activities (○) measured at 25°C. Data taken from [1,29].

change in the rate of at least one of the conformational changes making up the reaction cycle. However, there is another possibility, and that is that the reaction mechanism changes, with hydrophobic mismatch resulting in novel conformations for the protein never seen under 'normal' conditions. Indeed, it seems that this is the case. Figure 3 shows that the stoichiometry of Ca^{2+} binding to the Ca^{2+}-ATPase is only 1:1 in di($C_{14:1}$)PC, whereas it is 2:1 in di($C_{18:1}$)PC, as it is in the native membrane [1,7]. There is no evidence to suggest that states of the Ca^{2+}-ATPase with a single bound Ca^{2+} ion, capable of being phosphorylated by ATP, are formed in bilayers of di($C_{18:1}$)PC and thus it seems that a new conformational state of the Ca^{2+}-ATPase is formed in bilayers of di($C_{14:1}$)PC, not observed in bilayers of di($C_{18:1}$)PC. It is not known why a conformational state binding a single Ca^{2+} ion should be formed in di($C_{14:1}$)PC, but the Ca^{2+}-binding sites on the Ca^{2+}-ATPase are located between the transmembrane α-helices of the protein [8], and it is possible that changes in the packing of the helices, caused by a change in bilayer thickness, lead to changes in the Ca^{2+}-binding sites. A further point about the effects of fatty acyl chain length on the activity of the Ca^{2+}-ATPase is that the effects are highly co-operative, as shown by the data in Figure 3.

The results with the Ca^{2+}-ATPase show that effects of lipid structure on a membrane protein are likely to be complex and unique for each membrane protein. No simple model will describe adequately the effects that are possible. Rather, what is required is a detailed structural model for each protein, defining, in molecular detail, how lipids of particular structure interact with that particular membrane protein and what effects these interactions have on the structure and function of the protein.

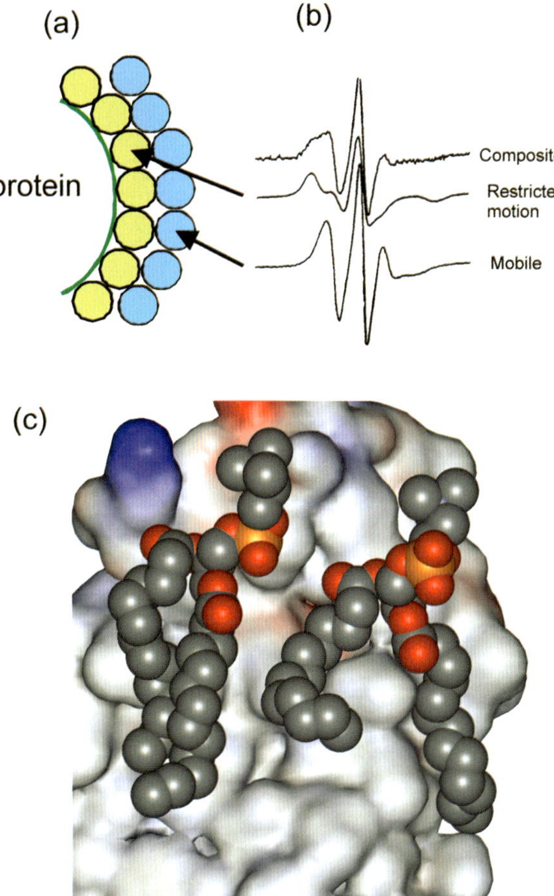

Figure 4. Annular lipid-binding sites
(**a**) A view of a membrane protein from the top, showing the first shell of lipid molecules (yellow) often referred to as the lipid annulus, and a second shell of lipid molecules (blue). (**b**) EPR spectra for a spin-labelled lipid in the presence of Ca^{2+}-ATPase. The experimental spectrum is a composite made up of a component due to lipid with restricted motion (the annular lipid) and a mobile component due to lipid not in contact with the protein. (**c**) Two of the annular lipid molecules bound on the surface of aquaporin are shown in space-fill format; the surface of the protein is shown coloured by charge (blue, positive; red, negative). Co-ordinates from PDB code 2B6O.

Annular lipid-binding sites

Water-soluble proteins are solvated by a layer of water molecules and the membrane-spanning region of a membrane protein is similarly solvated by a layer of lipid molecules. This was first shown using EPR spectroscopy with spin-labelled lipids [9]. The EPR spectrum of a spin-labelled lipid in the presence of a membrane protein such as the Ca^{2+}-ATPase from muscle sarcoplasmic reticulum shows two components (Figure 4) [10]. One component represents lipid molecules with restricted mobility, corresponding to the first shell of

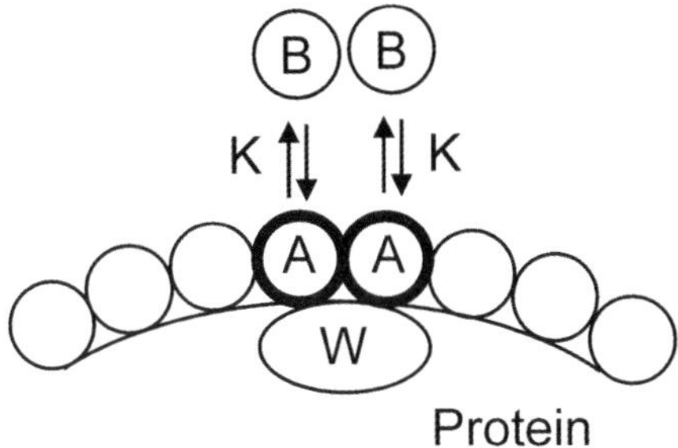

Figure 5. Lipid binding to an integral membrane protein
The hydrophobic surface of a membrane protein is covered by a set of lipid-binding sites, the annular or boundary sites. In this example, two of these sites, shown by the dark circles, are sufficiently close to the tryptophan residue shown that binding of a brominated lipid molecule to either of the sites will quench the fluorescence of the tryptophan residue. Therefore, in this case, $n = 2$.

lipid molecules interacting directly with the protein, the other component representing a mobile fraction of lipid molecules with a mobility similar to that of bulk lipid [9,10]. The first shell of lipid molecules is often referred to as the lipid annulus because it forms an annulus or ring around the protein [11]. This picture of lipid organization has now been confirmed by a variety of EM (electron microscopy) and X-ray diffraction studies that, for some membrane proteins, show examples of the annular lipid molecules, distorted by binding to the protein surface [12]: Figure 4 illustrates two such lipid molecules adopting highly distorted conformations on the surface of aquaporin [13].

Most of the lipid molecules in the annulus will be in rapid exchange with the bulk lipids in the membrane [10]. Of course, this fast exchange does not matter for the membrane protein; a lipid molecule will become distorted as soon as it moves on to the surface of the protein, and membrane proteins will always therefore be surrounded by a layer of distorted lipids, whatever the rate of lipid exchange.

We have developed fluorescence-quenching methods to measure lipid-binding constants at the annular lipid-binding sites. The approach assumes that the membrane-spanning region of a membrane protein is covered by a set of annular sites to which lipid molecules can bind (Figure 5). We wish to know whether a particular site has a preference for lipids of class X or for lipids of class Y. Binding of a molecule of lipid X or Y at each site around the protein can be described by a series of displacement reactions:

$$\text{Site.X} + \text{Y} \leftrightharpoons \text{Site.Y} + \text{X}$$

This equilibrium is described by an equilibrium constant K, the binding constant at the site for lipid Y relative to the binding constant at the site for lipid X, given by:

$$K = [\text{Site.Y}][\text{X}]/([\text{Site.X}][\text{Y}])$$

The aim is to measure values for K at the annular sites. The experimental approach makes use of the fact that the fluorescence of a tryptophan residue is quenched when the tryptophan residue is close to a bromine atom. The protein is reconstituted into bilayers containing a mixture of lipids X and Y where lipid X contains normal chains, whereas the chains in lipid Y contain two bromines per chain. If lipid Y binds close to a tryptophan residue in a membrane protein, then it will quench the fluorescence of the tryptophan residue. Because the time for two lipid molecules to exchange between the bulk lipid phase and the annular shell of lipids around a membrane protein is greater than the fluorescence lifetime [10,14], the level of fluorescence quenching observed in a mixture of lipids X and Y will be proportional to the fraction of annular sites occupied by lipid Y, and so will depend on the binding constant for lipid Y compared with that for lipid X. The higher the binding constant for lipid Y compared with that for lipid X, the more marked will be the fluorescence quenching; this is the basis for the measurement of binding constants. Determining numerical values for the relative binding constants also requires a knowledge of the number n of lipid-binding sites from which the fluorescence of a given tryptophan residues can be quenched (Figure 5); in practice, n is generally between 1 and 2 [15].

The method requires that the membrane protein being studied contains at least one tryptophan residue within the transmembrane region of the protein. Most membrane proteins, in fact, contain many tryptophan residues in the transmembrane region; although this aids sensitivity, it has the disadvantage that analysing the data generally requires the assumption that lipid-binding constants at sites close to all the tryptophan residues are the same, and this may not be true. An ideal membrane protein for study is therefore a protein lacking tryptophan residues, into which single tryptophan residues can be introduced using site-directed mutagenesis, allowing the determination of lipid-binding constants at a variety of positions around the circumference of the protein [16].

Lipid-binding constants have been measured using this approach as a function of lipid fatty acyl chain length [15,17–21]. For α-helical membrane proteins, the variation in binding constant with chain length is much less than expected from models based on the idea that the lipid distorts around a rigid membrane protein that does not change [4,22]. Rather, the cost of distorting the bilayer becomes so high for large stretchings or compressions that it becomes less costly to achieve hydrophobic matching by distorting the protein. These distortions could include changes in the tilt angles for the transmembrane α-helices, some unwinding of the α-helix and changes in the orientations of amino acid residues close to the lipid–water interface; such changes in the transmembrane α-helices could result in changes in the function of the membrane protein, as observed. For the β-barrel protein OmpF, agreement between the experimental data and lipid distortion models is much closer, suggesting that changing bilayer thickness has relatively little effect on rigid β-barrel proteins [4,23].

Studies with the mechanosensitive channel of large conductance, MscL, have shown that effects of lipid fatty acyl chain length on lipid-binding constants are different on the cytoplasmic and periplasmic sides of the membrane [21]. Effects of lipid headgroup structure on binding to MscL were also found to be different on the periplasmic and cytoplasmic sides [16]. In particular, whereas there was no selectivity in binding anionic and zwitterionic lipids on the periplasmic side, a hotspot binding anionic lipids with high affinity was found on the cytoplasmic side [16]. The hotspot consists of a cluster of three positively charged residues, adjacent in the primary sequence; binding of anionic lipid to the cluster modifies MscL function [24]. A similar positively charged cluster is seen in the potassium channel Kir2.2 and binding of anionic lipids such as PtdIns(4,5)P_2 to this cluster modulates channel function [25].

Non-annular lipid-binding sites

As well as the annular lipid molecules interacting with the bilayer-exposed surface of a membrane protein, lipid molecules are sometimes found buried within the protein, between transmembrane α-helices; these lipids have been referred to as non-annular lipids to distinguish them from the lipids bound to the annular sites [12]. These non-annular lipid molecules generally make contact with the annular lipid molecules and adopt an orientation much like that adopted by the lipid molecules in the bilayer, suggesting that non-annular lipid molecules incorporate into their binding sites by simple diffusion from the bilayer. Fluorescence-quenching methods can also be used to determine binding constants for lipids at these sites [26]. The homotetrameric potassium channel KcsA contains a set of non-annular binding sites, one at each protein–protein interface. The sites show a preference for binding anionic lipid, the binding constant for phosphatidylglycerol being ~3 times that for phosphatidylcholine [26]. Binding of anionic lipid to the site increases the open probability of the channel, probably by decreasing C-type inactivation [26].

Membrane protein structures in detergent

If the lipid bilayer is important in shaping a membrane protein, removing the lipid bilayer could lead to a large change in the structure of the protein. If this were the case, crystal structures of membrane proteins would only report distorted structures, as crystal structures are usually determined in the absence of lipid. In the case of the voltage-gated potassium channel K_vAP, the crystal structure determined in the absence of lipid is indeed highly distorted, but this is a rare example arising from the loose attachment of the voltage-sensor domain to the central core of the channel [27]. In the great majority of cases, it is clear that structures of membrane proteins determined from crystals grown from detergent micelles give a very good picture of the structure in the native membrane. The conclusion seems to be that the interactions between the transmembrane

α-helices of a membrane protein [28] are sufficiently specific to drive correct packing of the helices in the absence of a lipid bilayer. A good detergent will allow the transmembrane α-helices of a membrane protein to pack correctly, and at the same time, by interacting with the outer surface of the helix bundle, will prevent exposure of the hydrophobic helices to water or aggregation of the protein. An ideal detergent will have little structure of its own, allowing protein intramolecular interactions to determine the packing of the helices. In contrast, a lipid bilayer has a distinct three-dimensional structure, the most obvious feature of which is its hydrophobic thickness. As long as the hydrophobic thickness of the lipid bilayer in the absence of protein matches the hydrophobic thickness of the protein in the absence of the lipid bilayer, there is no problem and, in the membrane, the protein will be able to adopt its preferred structure. Membrane proteins have evolved so that the packing preferences of the membrane-spanning helices result in a structure that matches nicely that of the surrounding lipid bilayer, so that lipid and protein can meet without either having to change very much.

References

1. Starling, A.P., East, J.M. & Lee, A.G. (1993) Effects of phosphatidylcholine fatty acyl chain length on calcium binding and other functions of the (Ca^{2+}-Mg^{2+})-ATPase. *Biochemistry* **32**, 1593–1600
2. Froud, R.J., Earl, C.R.A., East, J.M. & Lee, A.G. (1986) Effects of lipid fatty acyl chain structure on the activity of the (Ca^{2+} + Mg^{2+})-ATPase. *Biochim. Biophys. Acta* **860**, 354–360
3. Froud, R.J., East, J.M., Jones, O.T. & Lee, A.G. (1986) Effects of lipids and long-chain alkyl derivatives on the activity of (Ca^{2+}-Mg^{2+})-ATPase. *Biochemistry* **25**, 7544–7552
4. Marsh, D. (2008) Energetics of hydrophobic matching in lipid–protein interactions. *Biophys. J.* **94**, 3996–4013
5. Starling, A.P., East, J.M. & Lee, A.G. (1995) Effects of phospholipid fatty acyl chain length on phosphorylation and dephosphorylation of the Ca^{2+}-ATPase. *Biochem. J.* **310**, 875–879
6. Starling, A.P., East, J.M. & Lee, A.G. (1996) Separate effects of long-chain phosphatidylcholines on dephosphorylation of the Ca^{2+}-ATPase and on Ca^{2+} binding. *Biochem. J.* **318**, 785–788
7. Michelangeli, F., Orlowski, S., Champeil, P., Grimes, E.A., East, J.M. & Lee, A.G. (1990) Effects of phospholipids on binding of calcium to (Ca^{2+}-Mg^{2+})-ATPase. *Biochemistry* **29**, 8307–8312
8. Moller, J.V., Olesen, C., Winther, A.M.L. & Nissen, P. (2010) The sarcoplasmic Ca^{2+}-ATPase: design of a perfect chemi-osmotic pump. *Q. Rev. Biophys.* **43**, 501–566
9. Marsh, D. (2008) Protein modulation of lipids, and vice versa, in membranes. *Biochim. Biophys. Acta* **1778**, 1545–1575
10. East, J.M., Melville, D. & Lee, A.G. (1985) Exchange rates and numbers of annular lipids for the calcium and magnesium ion dependent adenosinetriphosphatase. *Biochemistry* **24**, 2615–2623
11. Lee, A.G. (1977) Annular events: lipid–protein interactions. *Trends Biochem. Sci.* **2**, 231–233
12. Lee, A.G. (2003) Lipid–protein interactions in biological membranes: a structural perspective. *Biochim. Biophys. Acta* **1612**, 1–40
13. Gonen, T., Cheng, Y.F., Sliz, P., Hiroaki, Y., Fujiyoshi, Y., Harrison, S.C. & Walz, T. (2005) Lipid–protein interactions in double-layered two-dimensional AQPO crystals. *Nature* **438**, 633–638
14. London, E. & Feigenson, G.W. (1981) Fluorescence quenching in model membranes. 2. Determination of local lipid environment of the calcium adenosinetriphosphatase from sarcoplasmic reticulum. *Biochemistry* **20**, 1939–1948

15. Powl, A.M., East, J.M. & Lee, A.G. (2003) Lipid–protein interactions studied by introduction of a tryptophan residue: the mechanosensitive channel MscL. *Biochemistry* **42**, 14306–14317
16. Powl, A.M., East, J.M. & Lee, A.G. (2005) Heterogeneity in the binding of lipid molecules to the surface of a membrane protein: hot-spots for anionic lipids on the mechanosensitive channel of large conductance MscL and effects on conformation. *Biochemistry* **44**, 5873–5883
17. East, J.M. & Lee, A.G. (1982) Lipid selectivity of the calcium and magnesium ion dependent adenosinetriphosphatase, studied with fluorescence quenching by a brominated phospholipid. *Biochemistry* **21**, 4144–4151
18. Webb, R.J., East, J.M., Sharma, R.P. & Lee, A.G. (1998) Hydrophobic mismatch and the incorporation of peptides into lipid bilayers: a possible mechanism for retention in the Golgi. *Biochemistry* **37**, 673–679
19. Mall, S., Broadbridge, R., Sharma, R.P., Lee, A.G. & East, J.M. (2000) Effects of aromatic residues at the ends of transmembrane α-helices on helix interactions with lipid bilayers. *Biochemistry* **39**, 2071–2078
20. Williamson, I.M., Alvis, S.J., East, J.M. & Lee, A.G. (2002) Interactions of phospholipids with the potassium channel KcsA. *Biophys. J.* **83**, 2026–2038
21. Powl, A.M., East, J.M. & Lee, A.G. (2007) Different effects of lipid chain length on the two sides of a membrane and the lipid annulus of MscL. *Biophys. J.* **93**, 113–122
22. Lee, A.G. (2004) How lipids affect the activities of integral membrane proteins. *Biochim. Biophys. Acta* **1666**, 62–87
23. O'Keeffe, A.H., East, J.M. & Lee, A.G. (2000) Selectivity in lipid binding to the bacterial outer membrane protein OmpF. *Biophys. J.* **79**, 2066–2074
24. Powl, A.M., East, J.M. & Lee, A.G. (2008) Anionic phospholipids affect the rate and extent of flux through the mechanosensitive channel of large conductance MscL. *Biochemistry* **47**, 4317–4328
25. Tao, X., Avalos, J.L., Chen, J.Y. & Mackinnon, R. (2009) Crystal structure of the eukaryotic strong inward-rectifier K^+ Channel Kir2.2 at 3.1 Å resolution. *Science* **326**, 1668–1674
26. Marius, P., Zagnoni, M., Sandison, M.E., East, J.M., Morgan, H. & Lee, A.G. (2008) Binding of anionic lipids to at least three nonannular sites on the potassium channel KcsA is required for channel opening. *Biophys. J.* **94**, 1689–1698
27. Lee, S.Y., Lee, A., Chen, J.Y. & Mackinnon, R. (2005) Structure of the KvAP voltage-dependent K^+ channel and its dependence on the lipid membrane. *Proc. Natl. Acad. Sci. U.S.A.* **102**, 15441–15446
28. Popot, J.L. & Engelman, D.M. (2000) Helical membrane protein folding, stability, and evolution. *Annu. Rev. Biochem.* **69**, 881–922
29. Lee, A.G. (1998) How lipids interact with an intrinsic membrane protein: the case of the calcium pump. *Biochim. Biophys. Acta* **1376**, 381–390

Biochem. Soc. Symp. 78
Citation reference: Biochem. Soc. Trans. (2011) **39**, 767–774.

8

Lipid–protein interactions as determinants of membrane protein structure and function

William Dowhan[1] and Mikhail Bogdanov

Department of Biochemistry and Molecular Biology, University of Texas Medical School-Houston, 6431 Fannin Street, Houston, TX 77030, U.S.A.

Abstract

To determine how the lipid environment affects membrane protein structure and function, strains of *Escherichia coli* were developed in which normal phospholipid composition can be altered or foreign lipids can be introduced. The properties of LacY (lactose permease) were investigated as a function of lipid environment. Assembly of LacY in membranes lacking PE (phosphatidylethanolamine) results in misorientation of the N-terminal six-TM (transmembrane domain) helical bundle with loss of energy-dependent uphill transport and retention of energy-independent downhill transport. Post-assembly introduction of PE results in nearly native orientation of TMs and restoration of uphill transport. Foreign lipids with no net charge can substitute for PE in supporting native LacY topology, but restoration of uphill transport is dependent on native topology and the proper folding of a solvent-exposed domain. Increasing the positive charge density of the cytoplasmically exposed surface of LacY counters TM misorientation in the absence of neutral lipids, demonstrating that charge interactions between these domains and

[1]*To whom correspondence should be addressed (email William.Dowhan@uth.tmc.edu).*

the surface of the membrane bilayer are determinants of TM orientation. Therefore membrane protein organization or reorganization is determined either during initial assembly or post-insertionally through direct interactions between the protein and the lipid environment, which affects the topogenic potency of opposing charged residues as topological signals independent of the translocon.

Introduction

The process by which polytopic membrane proteins are initially inserted into the membrane bilayer has been reviewed extensively [1,2]. Final orientation of TMs (transmembrane domains) and folding of membrane proteins is determined by the membrane-protein-insertion machinery (translocon) [3,4], topogenic sequences within the protein [5–7] and the properties of the lipid bilayer [5,8–10]. The primary driving force for membrane insertion of TMs is their hydrophobicity [11–13]. The orientation of TMs with respect to the plane of the lipid bilayer is largely determined by the positive-inside rule [14], which is based on the observed bias for exposure of positively charged extramembrane domains to the cytoplasm with neutral or negatively charged extra-membrane domains oriented to the opposite side of the membrane. However, molecular genetic manipulation of membrane lipid composition has revealed a role for the properties of the lipid bilayer in determining TM orientation of polytopic membrane proteins [15]. Although gross lipid composition of individual membranes may not change dramatically, local changes of lipid composition within a given membrane do occur. In eukaryotic cells, lipid composition varies greatly between different membranes within the same cell [16], which exposes proteins to a varying lipid environment during intracellular trafficking. Understanding the effects of membrane lipid composition even in membranes where composition may not vary is of importance, since topogenic signals have evolved in the context of the host lipid composition [15], which varies between different organisms.

In the present paper, the molecular genetic manipulation of *Escherichia coli* membrane lipid composition is reviewed. The effects of changes in protein sequence and changes in membrane lipid composition on polytopic membrane protein TM orientation are integrated. The results strongly indicate that protein sequence and membrane lipid composition have co-evolved as synergistic determinants of membrane protein organization.

Genetic manipulation of membrane lipid composition

Surprisingly, large changes in *E. coli* membrane lipid composition are not lethal, even though there are significant changes in cell phenotype (see [17] for a detailed review and references therein). The major glycerophosphate-based lipids (phospholipids) of *E. coli* are zwitterionic PE (phosphatidylethanolamine)

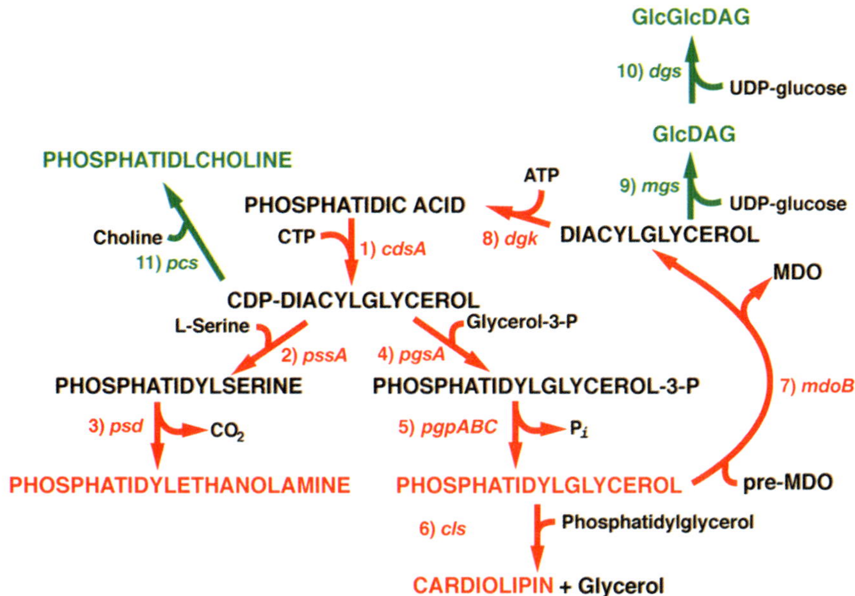

Figure 1. Synthesis of native and foreign lipids in *E. coli*
Pathways, genes encoding the enzyme responsible and major phospholipids naturally occurring in *E. coli* are indicated in red. Pathways, genes encoding the enzyme responsible and the major lipid synthesized by foreign enzymes in *E. coli* are indicated in green. 1. *cdsA*, CDP-diacylglycerol synthase; 2. *pssA*, phosphatidylserine synthase; 3. *psd*, phosphatidylserine decarboxylase; 4. *pgsA*, phosphatidylglycerophosphate synthase; 5. *pgpABC*, phosphatidylglycerophosphate phosphatase (any of three gene products catalyse this step [18]); 6. *cls*, CL synthase; 7. *mdoB*, PG:pre-MDO (membrane-derived oligosaccharide) *sn*-glycerol-1-phosphate transferase; 8. *dgk*, diacylglycerol kinase; 9. *mgs*, GlcDAG synthase (*Acholeplasma laidlawii*); 10. *dgs*, GlcGlcDAG synthase (*Acholeplasma laidlawii*); 11. *pcs*, PC synthase (*Legionella pneumophila*). Modified from [9] with permission. © 2009 The American Society for Biochemistry and Molecular Biology.

(~70 %), anionic PG (phosphatidylglycerol) (~20 %) and CL (cardiolipin; diphosphatidylglycerol) (~5 %) (Figure 1). The remaining phospholipids are less than 5 % of the total pool. The ratio of zwitterionic to anionic phospholipids can be varied by mutations in the *pssA* and *pgsA* genes. Null mutations in the *pgsA* gene are lethal, but when combined with a set of suppressor mutations, viable cells can be grown with ~95 % PE and ~5 % primarily phosphatidic acid and CDP-diacylglycerol. Null mutations in the *pssA* gene result in complete elimination of all amino-containing and zwitterionic phospholipids, with 100 % of phospholipids being anionic. Both genes have been placed under the control of inducible promoters so that the ratio of zwitterionic to anionic phospholipids can be systematically varied. Cells with mutations in *pgsA* have been used to establish a role for anionic lipids as sites for organization of integral membrane proteins with amphitropic proteins to form membrane associated molecular machines responsible for cell division, DNA replication, protein secretion and membrane protein insertion.

Cells completely lacking PE (*pssA*-null strains) are dependent on millimolar levels of bivalent cations ($Ca^{2+} > Sr^{2+} > Mg^{2+}$, with Ba^{2+} being ineffective) to prevent cell lysis and possibly to support non-bilayer properties of CL; PE is the predominant phospholipid with non-bilayer properties in wild-type cells. Lack of PE also compromises late stages of cell division and final constriction both in *E. coli* and eukaryotic cells. In addition, PE is required to support the energy-dependent uphill transport of solutes catalysed by several amino acid and sugar permeases. Detailed analysis of the molecular basis for the lack of uphill transport function discussed below uncovered the involvement of membrane lipid composition as a determinant of topological organization of some membrane proteins and further extended the rules governing membrane protein folding and assembly.

E. coli membrane lipid composition can be further manipulated by introduction of foreign genes [9] that encode lipids not found in *E. coli* (Figure 1). Expression of such genes in cells lacking PE allows the study of the effects of lipids with a different, but overlapping, mixture of physical and chemical properties when compared with PE. PE shows non-bilayer-forming properties dependent on fatty acid composition, is charged but carries no net charge, and is capable of hydrogen-bonding through its amine-containing headgroup. PG and CL carry a net negative charge and have headgroup hydrogen-bonding ability, but PG is bilayer-prone whereas CL is non-bilayer prone in the presence of bivalent cations, which are prevalent in cells. Introduction of the *pcs* gene results in replacement of PE by 70 % of total phospholipid as PC (phosphatidylcholine). PC is similar to PE in being net-charge-neutral, but is a bilayer-prone lipid. However, the quaternary amine of PC cannot hydrogen-bond. Introduction of the *mgs* and *dgs* genes into PE-lacking cells results in 40 % of the total glycerol-based lipids as either GlcDAG (monoglucosyl diacylglycerol) or GlcGlcDAG (diglucosyl diacylglycerol) respectively. Both carry no charge, have hydrogen-bonding properties, and, like PE and PC, can dilute the high negative charge density of the membrane surface resulting from PG and CL. However, they differ in that GlcDAG is non-bilayer-prone, whereas GlcGlcDAG is bilayer-prone.

Lipid effects on folding of an extramembrane domain that is crucial for uphill transport function

On the basis of *in vitro* reconstitution of LacY (lactose permease) of *E. coli* into proteoliposomes, it has been recognized for over 35 years that PE is required for energy-dependent uphill accumulation of substrate, but not for energy-independent downhill equilibration of substrate [19]. Unexpectedly, PC does not substitute for PE in supporting uphill transport by LacY in proteoliposomes [19,20]. However, phosphatidylserine did substitute, suggesting a requirement for an ionizable amine to support uphill transport. With the construction of strains of *E. coli* with different lipid compositions, the physiological significance of lipid requirements for transport function could be tested *in vivo*. Using strains of *E. coli* either with or without PE, the requirement for PE for uphill transport and not for downhill transport was verified *in vivo* [21]. Surprisingly, cells with PC in place of PE also supported uphill transport [8]. Strains containing GlcDAG [22] or phosphatidylserine [23] in place of PE also supported uphill transport. However, GlcGlcDAG did not support uphill transport [24]. Part of the answer for the above lipid requirements was obtained using a monoclonal antibody (mAb 4B1) specific for the sequence and conformation of an epitope in the extramembrane domain P7 of LacY [25] that is exposed to the periplasm (Figure 2A). This antibody recognizes LacY in membranes and after SDS/PAGE and Western blotting, provided LacY initially exhibits uphill transport function [26,27]. Perturbation of P7 domain organization results in a lowering of the abnormally high pK_a of a glutamate residue residing within TM X [25]. The monoclonal antibody recognizes LacY from PE-containing cells or from cells in which LacY was originally assembled in the absence of PE and then exposed to PE either *in vitro* during refolding on a solid support [26,27], *in situ* during or after insertion into isolated *E. coli* inside-out membrane vesicles [28] or *in vivo* after assembly [5,27] by interaction with non-native late-folding intermediates of LacY, like most conventional protein molecular chaperones [29]. Recognition was also observed for LacY from cells containing PC or GlcDAG, but not GlcGlcDAG [8], in place of PE. LacY was not recognized from cells containing only PG and CL [26,27]. The PC species used in most reconstitutions of LacY were those with high unsaturated fatty acid content [19,20], whereas *E. coli*-derived PC [8,19] is a mixture of species containing significant amounts of saturated fatty acids. Proper folding of the P7 domain requires PC species with at least one saturated fatty acid in combination with anionic lipid possessing hydrogen-bonding capability [8]. Therefore support of uphill transport function and recognition by mAb 4B1 appears to require a complex mixture of physical and chemical properties dictated by the hydrophilic headgroup and the hydrophobic backbone of the supporting lipid.

Lipids as topological determinants

The molecular basis for lack of uphill transport by LacY in cells containing only PG and CL is due to a large topological inversion that also disrupts the structure of the P7 domain [5,27]. The SCAM^{TM} (substituted cysteine accessibility method as applied to TM orientation) was used to map the organization of LacY as a function of membrane lipid composition. In this method [30], single cysteine replacements were made in a LacY derivative lacking normal cysteine residues. After synthesis in cells of various lipid compositions, the accessibility of the cysteine residues to a membrane-impermeable

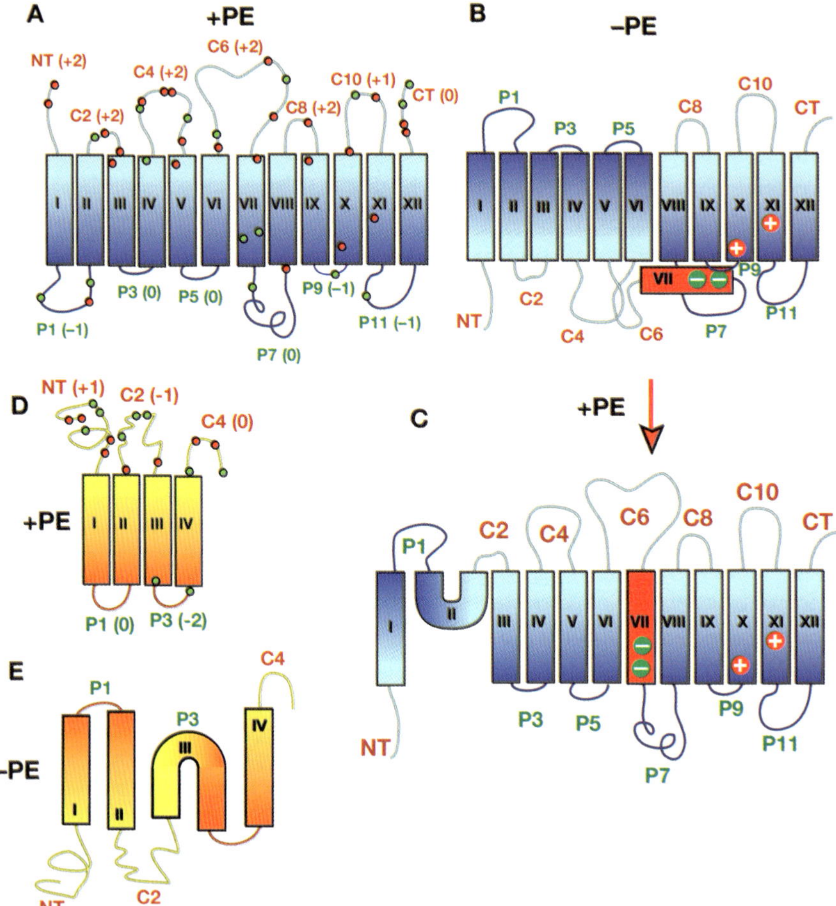

Figure 2. TM organization of secondary transporters as a function of membrane lipid composition

The top of each structure faces the cytoplasm. Blue, red and yellow/orange rectangles denote TMs numbered sequentially by Roman numerals. N- and C-terminal domains are labelled as NT and CT respectively. Extramembrane domains are labelled according to their exposure to the cytoplasm (C) or periplasm (P) in PE-containing cells. (**A**) Topology organization of LacY as determined in PE-containing (+PE) cells is depicted. The net charge of each extramembrane domain is noted next to the domain name, and the approximate positions of charged residues throughout the protein are indicated by green (negative charge) and red (positive charge) dots. (**B**) Topology organization of LacY as determined in PE-lacking (−PE) cells is shown. TM VII (red) exposure to the periplasm results in the loss of salt bridges between the two aspartate residues in TM VII with positively charged amino acids in TM X and TM XI. (**C**) Topology organization of LacY as determined in cells where assembly of LacY initially occurred in the absence of PE followed by the induction of PE synthesis in the absence of new LacY synthesis. (**D**) Topology organization of PheP (topology of lipid-insensitive region TM V-CT not shown) as determined in PE-containing cells is shown with labelling of domains following that described in (**A**). (**E**) Topology organization of PheP as determined in cells lacking PE. Note that the topological organization of GabP is the same as that of PheP with a slightly different distribution of charged residues. Modified from [9] with permission. © 2009 The American Society for Biochemistry and Molecular Biology.

thiol-group-specific reagent linked to biotin was determined in whole cells or lysed cells. Cysteine residues in periplasmic domains should be accessible in whole cells and those in cytoplasmic domains should be accessible only after cell lysis, whereas those in TMs should not be accessible under either condition since the thiol-specific reagent reacts only with ionized cysteine residues in a hydrophilic environment. The method was extended further to differentiate locally restricted domains from TMs or domains that partially insert into the membrane without traversing the membrane (mini-loops) [5,30]. Use of mild alkaline treatment during derivatization can disrupt local structurally hindered cysteine residues within solvent-exposed domains, whereas strong alkaline treatment can release mini-loops from the membrane without exposing cysteine residues within TMs.

Assembly of LacY in PE-containing cells results in the expected orientation of all TMs [31] on the basis of all previous biochemical and structural information on LacY (Figure 2A). However, assembly in cells lacking PE (Figure 2B) results in the complete inversion of the N-terminal six-TM helical bundle with respect to the plane of the membrane bilayer and the last five TMs [5,27]. In addition TM VII no longer resides in the membrane, but exists as an extramembrane domain exposed to the periplasm [5]. TM VII is of low hydrophobicity due to two aspartate residues that are normally salt-bridged to positive residues in TMs X and XI [32,33]. Remarkably, downhill transport is not affected by this topological inversion, probably because the N-terminal bundle, with most of the substrate-binding sites [32], remains intact. However, this inversion results in a major disruption of secondary structure in the vicinity of domain P7, which contains the recognition site for mAb 4B1. Reconstitution of LacY isolated either from PE-containing or PE-lacking cells into proteoliposomes results in a topological organization dictated by the lipid composition of the proteoliposomes and not the lipid composition of the source of LacY [20]. Although PC species with high unsaturated fatty acid content used in proteoliposomes does not support uphill transport or P7 conformation [19,20], they do support wild-type topology as does total *E. coli* total phospholipid composition [20]. However, presence of PG and CL alone in liposomes results an inverted topology [20]. *In vivo* substitution of PE with similarly net-neutral PC [8], GlcDAG [34] or GlcGlcDAG [24] also supports wild-type topology. Therefore the relevant property of lipids that support wild-type topological organization is their net-charge-neutral headgroup character that dilutes out the high negative charge density of the membrane surface contributed by PG and CL [15].

Most remarkable is that topological organization and TM orientation, once attained, is dynamic and can change in response to changes in the lipid environment [5,27]. Using an inducer-regulated promoter for *pssA* gene expression, LacY was synthesized and fully assembled in the absence of PE. After termination of new LacY synthesis, PE synthesis was induced with PE returning to wild-type levels. Post-assembly synthesis of PE resulted in a regain of recognition by mAb 4B1 and uphill transport function, insertion of TM VII into the membrane, and a near complete return to native topological organization [5]. Only NT, TM I, P2 and TM II did not return to native organization (Figure 2C).

Properties of lipid-sensitive domains

The topological organization of amino acid permeases for phenylalanine (PheP) [35] and γ-aminobutyrate (GabP) [36] are also sensitive to the lipid composition (Figures 2D and 2E). In PE-lacking cells, the N-terminal TM hairpin of these two transporters is inverted with respect to the remainder of the protein and TM III exists as a mini-loop. This inversion also results in loss of uphill transport function. In the case of PheP, wild-type topological organization and function were restored by post-assembly synthesis of PE. For the amino acid permeases, the N-terminal extramembrane domains do not follow the positive-inside rule and are, in some cases, net negative on the cytoplasmic side (Figure 2D). Although the extramembrane domains of the N-terminal bundle of LacY follow the positive-inside rule, there are more negatively charged amino acids on the cytoplasmic surface of the N-terminal than the C-terminal half of LacY (Figure 2A).

To test whether the presence of negative residues in these domains makes the protein topology dependent on lipid composition, the mixture of charged residues on the cytoplasmic surface of the N-terminal six-TM bundle of LacY was altered, and topological organization as a function of lipid composition was determined. Increasing the net positive charge of the bundle surface by 1 in a position-independent manner (i.e. within C-3, C-4 or C-6), either by adding a positive charge or removing a negative charge, prevented inversion in PE-lacking cells [5]. Adding a positive and negative charge (no net change) did not prevent inversion. If increasing the positive charge of the N-terminal bundle prevents inversion in PE-lacking cells, then increasing the negative charge should induce inversion in PE-containing cells. This proved to be the case, but inversion only occurred after changing the net charge of the cytoplasmic surface of the bundle from +6 to −6. Therefore the main effect of net-neutral lipids is to dampen the translocation potential of negative residues in opposition to the positive-inside rule to allow the presence of negative residues in the cytoplasmic domain for functional purposes without affecting protein topology. This conclusion was supported further by results with PheP and CscB (sucrose permease) [37]. Topological inversion for PheP is prevented in PE-lacking cells by increasing the net positive charge in domains NT and C2. The topology of wild-type CscB is not sensitive to lipid composition. However, decreasing the net positive charges within the cytoplasmic domains of the N-terminal six-TM bundle results in inversion of the bundle first in PE-lacking cells and, upon further decreases in positive charges, inversion occurs in PE-containing cells. These results are consistent with previous results that indicate dominance for positive charge over negative charge as a topological determinant, but now demonstrate that the translocation potential of negative residues is greatly enhanced in the absence of PE [5].

PE and the positive-inside rule

Although the positive-inside rule is generally accepted [14,38], the cellular factors involved in its execution and how positively charged residues exert their effect on topology are not well established. The contribution of the translocon to making a topological decision is limited by the size of newly synthesized proteins, the time of protein residency within the translocon and the effective size of the translocation pore, which is still a matter of debate [4,39]. Since orientation of TMs within a membrane protein can be changed by the addition or removal of a single positively charged residue or by the introduction of one or many negatively charged residues depending on the absence or presence of PE respectively, the question of the relative topological potency of the charged residues in the context of different phospholipid compositions is raised. The translocation block provided by protonated positively charged amino acid residues within cytoplasmic segments serves as a primary factor determining TM topology; however, the effective charge of these domains can be affected by mechanisms that selectively protonate or deprotonate negatively charged amino acid residues depending of the presence or absence of PE [9,40]. Whatever the precise mechanism of protonation/deprotonation of these residues, PE and other neutral phospholipids clearly attenuate the topological effects of acidic residues [15]. Negative residues in the absence of PE exert an increased translocation potential that results in translocation of domains that now exhibit a lower effective net positive charge. The retention potential of positively charged residues is enhanced in the presence of PE or other net-neutral lipids by reducing the translocation potential of negative residues, thus providing a molecular basis for the operation of the positive-inside rule for domains containing a mixture of positive and negative residues. LacY and PheP initially misoriented by assembly in membranes lacking PE can be properly oriented by post-assembly exposure to PE due to strengthening of the positive retention potential to drive reorientation of TMs.

The membrane potential (outside-positive in *E. coli*) may actively promote translocation of acidic residues and impede translocation of basic residues and therefore contribute to the establishment of topology governed by the positive-inside rule [41,42]. However, in the endoplasmic reticulum membrane, where no detectable or defined membrane potential besides a Ca^{2+} gradient is found, proteins still follow the positive-inside rule [11]. The positive-inside rule also applies to obligate acidophiles, such as *Sulfolobus acidocaldarius*, where membrane potential is permanently reversed (i.e. inside-positive instead of negative), suggesting that the positive-inside rule may not be dependent on the polarity of the membrane potential [43]. Therefore the net effect of positive charge of cytoplasmic domains as determined by the membrane lipid composition may be dominant in retention of the positively charged residues on the cytoplasmic side of different membrane systems.

Effect of flexible domains between differentially lipid-sensitive domains

In the three proteins (LacY, PheP and GabP) demonstrated to date to contain a mixture of domains whose topologies are sensitive and insensitive to the lipid environment, a TM (Figures 2B, 2C and 2E) exists that apparently acts as a flexible molecular hinge between the lipid-sensitive and -insensitive domains. For LacY, TM VII exits the membrane in PE-lacking cells and TM II forms a mini-loop during reorientation of LacY after post-assembly introduction of PE [5]. TM III of PheP and GabP forms a mini-loop in PE-lacking cells [35,36]. Are such connecting flexible domains required in order for the flanking domains to respond differently to the lipid environment? This was tested for LacY by increasing the hydrophobicity of TM VII of LacY through replacement of Asp^{240} with isoleucine. This substitution prevented inversion and solvent-exposure of TM VII of wild-type LacY in cells lacking PE as well as the LacY derivative with a net -6 charge for its cytoplasmic surface in PE-containing cells [5]. Therefore the unfavourable energetics of solvent-exposure of a hydrophobic domain can override the favourable energetics of inversion of the N-terminal bundle. This may explain why the majority of proteins are not sensitive to membrane lipid composition. Therefore lipid-dependent protein folding is dependent on both short-range interactions of charged extramembrane domains with the lipid environment and long-range co-operative interactions of TM helix packing. Unfortunately, predictions are difficult to make of domains that might act as flexible molecular switches allowing independent responses to the lipid environment.

Where is the final TM organization established?

Although the initial orientation of TM domains may be governed by interactions between the nascent polypeptide chain and the translocon, the final TM orientation depends on short-range interactions between the protein and the lipid environment and long-range interactions within the protein during final folding events [15]. The topogenic influence of lipids appears to be largely independent of other protein factors, including the translocon, since topological organization and transport function of LacY reconstituted into protein-free liposomes are determined solely by the phospholipid composition of the liposome independent of the lipid composition of the cells from which LacY was derived [20]. During reorientation of misoriented LacY by post-assembly introduction of PE, it is unlikely that the translocon is involved; however, the involvement of other proteins such as chaperones has not been ruled out. LacY forms a compact folded structure in PE-lacking cells on the basis of its function in downhill transport [21,31], and it must therefore exist as a structure free of association with the translocon. Utilization of the translocon would require reassociation and unfolding to effect reorientation. LacY is also in large molar excess, when overproduced, over the number of functional translocons. During initial synthesis and assembly of LacY, the hydrophobicity of TM VII and the charge nature of its preceding cytoplasmic domain determine final topology [5]. Since it is unlikely that the translocon can accommodate even five TMs in its aqueous environment [1], much of the N-terminal bundle of LacY must exist in a topologically undecided state outside the translocon until TM VII is synthesized [5]. Therefore all of these results are consistent with final topological decisions being made outside and independent of the translocon and dictated largely by lipid–protein interactions and the positive-inside rule as applied to a multiple TM unit.

Conclusions and perspectives

By artificially changing the steady-state membrane lipid composition, new principles governing membrane protein assembly have been uncovered. In addition, membrane protein structure during assembly and after assembly has been shown to be highly dynamic and not fixed or static. Changes in the charge nature of the extramembrane domains connecting TMs or changes in the net charge of the membrane surface synergistically can result in similar changes in the topological organization of a membrane protein, thus supporting charge interaction in a complementary manner between the cytoplasmic domains and the bilayer surface as a determinant of topological organization. Therefore proteins and lipids have co-evolved to establish a set of interdependent determinants of protein organization. Zwitterionic or net-neutral lipids are required to fulfil the positive-inside rule, which explains why positive residues are more potent topological determinants than negative residues under physiological conditions. During protein translation, LacY exists in a topologically undetermined state until at least TM VII has exited the ribosome. This is supported by the fact that the final topology of the N-terminal six-TM bundle is dependent on both the charge nature of C-6 and the hydrophobicity of TM VII. This fact disfavours a linear or sequential mode of TM insertion and demonstrates that the positive-inside rule can be executed in a retrograde manner and that long-range intramolecular interactions can influence final protein folding events. Subsequent to these findings, a similar conclusion was reached for EmrE of *E. coli* [44]. The results also definitively and systematically demonstrate that LacY and PheP can undergo TM flipping or changes in orientation after folding into a compact structure and that topology of the N-terminal TM bundle of LacY and the N-terminal TM hairpin of PheP or GabP are most probably determined outside the translocon. Therefore membrane protein organization is determined during initial assembly through interactions between the protein and the lipid environment, but is also dynamic and can change post-assembly, dependent on changes in the lipid environment. LacY, GabP and PheP are the best established examples of this dynamic aspect of native polytopic membrane proteins.

The fact that changes in lipid environment can have such a dramatic effect on membrane protein organization has important implications for membrane proteins in eukaryotic systems or in membranes where lipid composition is not uniform either spatially or temporally. Eukaryotic membrane proteins are exposed to various lipid environments as they move along the secretory pathway from the endoplasmic reticulum to the Golgi and finally to either the plasma membrane or other internal organelles. This affords the opportunity for initial latent activity to be activated at the site of final protein residency by the proper lipid environment. Local changes in lipid environment through interaction with specifically enriched lipid domains, lipid rafts and membrane fission or fusion sites could also activate or inactivate proteins. Lipid–protein interactions during polytopic protein biogenesis can contribute to folding anomalies induced by minor sequence perturbations in inherited topological disorders. Naturally occurring mutations of structurally important residues could lead to a different structural organization of the mutant protein within the changing lipid environment as the protein moves through the same or altered organelle trafficking route. Therefore changes in membrane lipid composition either locally or during intracellular movement of proteins along the organelle-based secretory pathway can lead to misinterpretation of existing topological signals or different interpretation of new ones. Incompatibility of lipids necessary for correct topogenesis and misinterpretation of topological signals by the 'wrong' lipid profile may be the basis for the inability to obtain active membrane proteins after expression in foreign host cells.

Clearly not all membrane proteins are sensitive to the lipid environment, since cells with large changes in membrane lipid composition are viable. Even in the case of solute permeases, partial transport function is retained in the absence of PE. Other functions, such as cell division, are also impaired in the absence of PE. The rescue of LacY function *in vivo* by PC [8] and GlcDAG [34] does not result in a completely wild-type structure as is evident from altered accessibility of some cysteine substitutions in the extramembrane domains. The availability of an extensive collection of mutants in which membrane lipid composition can be systematically controlled provides an important resource for the study of the physiologically relevant role of lipids in a wide range of biological processes. Since membrane protein sequence is 'written' for a given membrane environment and some topogenic signals may dominate over others within different lipid profiles, it is tempting to speculate that, during the course of evolution, both proteins and lipids co-evolved in the context of the lipid environment of membrane systems in which both are mutually dependent on each other.

Funding

This work was supported in whole or in part by the National Institutes of General Medical Sciences [grant number GM R37 20478] and the John Dunn Research Foundation (to W.D.)

References

1. Driessen, A.J. & Nouwen, N. (2008) Protein Translocation across the bacterial cytoplasmic membrane. *Annu. Rev. Biochem.* **77**, 643–667
2. Rapoport, T.A. (2007) Protein translocation across the eukaryotic endoplasmic reticulum and bacterial plasma membranes. *Nature* **450**, 663–669
3. Junne, T., Kocik, L. & Spiess, M. (2010) The hydrophobic core of the Sec61 translocon defines the hydrophobicity threshold for membrane integration. *Mol. Biol. Cell* **21**, 1662–1670
4. Skach, W.R. (2009) Cellular mechanisms of membrane protein folding. *Nat. Struct. Mol. Biol.* **16**, 606–612
5. Bogdanov, M., Xie, J., Heacock, P. & Dowhan, W. (2008) To flip or not to flip: protein–lipid charge interactions are a determinant of final membrane protein topology. *J. Cell Biol.* **182**, 925–935
6. Monne, M., Hessa, T., Thissen, L. & von Heijne, G. (2005) Competition between neighboring topogenic signals during membrane protein insertion into the ER. *FEBS J.* **272**, 28–36
7. von Heijne, G. & Gavel, Y. (1988) Topogenic signals in integral membrane proteins. *Eur. J. Biochem.* **174**, 671–678
8. Bogdanov, M., Heacock, P., Guan, Z. & Dowhan, W. (2010) Plasticity of lipid–protein interactions in the function and topogenesis of the membrane protein lactose permease from *Escherichia coli*. *Proc. Natl. Acad. Sci. U.S.A.* **107**, 15057–15062
9. Bogdanov, M., Jun, X. & Dowhan, W. (2009) Lipid–protein interactions drive membrane protein topogenesis in accordance with the positive-inside rule. *J. Biol. Chem.* **284**, 9637–9641
10. White, S.H. & von Heijne, G. (2005) Do protein–lipid interactions determine the recognition of transmembrane helices at the ER translocon? *Biochem. Soc. Trans.* **33**, 1012–1015
11. Harley, C.A., Holt, J.A., Turner, R. & Tipper, D.J. (1998) Transmembrane protein insertion orientation in yeast depends on the charge difference across transmembrane segments, their total hydrophobicity, and its distribution. *J. Biol. Chem.* **273**, 24963–24971
12. Lee, E. & Manoil, C. (1994) Mutations eliminating the protein export function of a membrane-spanning sequence. *J. Biol. Chem.* **269**, 28822–28828
13. White, S.H., Ladokhin, A.S., Jayasinghe, S. & Hristova, K. (2001) How membranes shape protein structure. *J. Biol. Chem.* **276**, 32395–32398
14. von Heijne, G. (1986) The distribution of positively charged residues in bacterial inner membrane proteins correlates with the trans-membrane topology. *EMBO J.* **5**, 3021–3027
15. Dowhan, W. & Bogdanov, M. (2009) Lipid-dependent membrane protein topogenesis. *Annu. Rev. Biochem.* **78**, 515–540
16. Dowhan, W. (1997) Molecular basis for membrane phospholipid diversity: why are there so many lipids? *Annu. Rev. Biochem.* **66**, 199–232
17. Dowhan, W. (2009) Molecular genetic approaches to defining lipid function. *J. Lipid Res.* **50**, S305–S310
18. Lu, Y.-H., Guan, Z., Zhao, J. & Raetz, C.R.H. (2011) Three phosphatidylglycerol-phosphate phosphatases in the inner membrane of *Escherichia coli*. *J. Biol. Chem.* **286**, 5506–5518
19. Chen, C.C. & Wilson, T.H. (1984) The phospholipid requirement for activity of the lactose carrier of *Escherichia coli*. *J. Biol. Chem.* **259**, 10150–10158
20. Wang, X., Bogdanov, M. & Dowhan, W. (2002) Topology of polytopic membrane protein subdomains is dictated by membrane phospholipid composition. *EMBO J.* **21**, 5673–5681

21. Bogdanov, M. & Dowhan, W. (1995) Phosphatidylethanolamine is required for *in vivo* function of the membrane-associated lactose permease of *Escherichia coli*. *J. Biol. Chem.* **270**, 732–739
22. Wikstrom, M., Xie, J., Bogdanov, M., Mileykovskaya, E., Heacock, P., Wieslander, A. & Dowhan, W. (2004) Monoglucosyldiacylglycerol, a foreign lipid, can substitute for phosphatidylethanolamine in essential membrane-associated functions in *Escherichia coli*. *J. Biol. Chem.* **279**, 10484–10493
23. Hawrot, E. & Kennedy, E.P. (1978) Phospholipid composition and membrane function in phosphatidylserine decarboxylase mutants of *Escherichia coli*. *J. Biol. Chem.* **253**, 8213–8220
24. Wikstrom, M., Kelly, A., Georgiev, A., Eriksson, H., Rosen-Klement, M., Bogdanov, M., Dowhan, W. & Wieslander, A. (2009) Lipid-engineered *Escherichia coli* membranes reveal critical lipid head-group size for protein function. *J. Biol. Chem.* **284**, 954–965
25. Sun, J., Wu, J., Carrasco, N. & Kaback, H.R. (1996) Identification of the epitope for monoclonal antibody 4B1 which uncouples lactose and proton translocation in the lactose permease of *Escherichia coli*. *Biochemistry* **35**, 990–998
26. Bogdanov, M., Sun, J., Kaback, H.R. & Dowhan, W. (1996) A phospholipid acts as a chaperone in assembly of a membrane transport protein. *J. Biol. Chem.* **271**, 11615–11618
27. Bogdanov, M., Umeda, M. & Dowhan, W. (1999) Phospholipid-assisted refolding of an integral membrane protein: minimum structural features for phosphatidylethanolamine to act as a molecular chaperone. *J. Biol. Chem.* **274**, 12339–12345
28. Bogdanov, M. & Dowhan, W. (1998) Phospholipid-assisted protein folding: phosphatidylethanolamine is required at a late step of the conformational maturation of the polytopic membrane protein lactose permease. *EMBO J.* **17**, 5255–5264
29. Bogdanov, M. & Dowhan, W. (1999) Lipid-assisted protein folding. *J. Biol. Chem.* **274**, 36827–36830
30. Bogdanov, M., Zhang, W., Xie, J. & Dowhan, W. (2005) Transmembrane protein topology mapping by the substituted cysteine accessibility method (SCAM^{TM}): application to lipid-specific membrane protein topogenesis. *Methods* **36**, 148–171
31. Bogdanov, M., Heacock, P.N. & Dowhan, W. (2002) A polytopic membrane protein displays a reversible topology dependent on membrane lipid composition. *EMBO J.* **21**, 2107–2116
32. Abramson, J., Iwata, S. & Kaback, H.R. (2004) Lactose permease as a paradigm for membrane transport proteins. *Mol. Membr. Biol.* **21**, 227–236
33. Abramson, J., Smirnova, I., Kasho, V., Verner, G., Kaback, H.R. & Iwata, S. (2003) Structure and mechanism of the lactose permease of *Escherichia coli*. *Science* **301**, 610–615
34. Xie, J., Bogdanov, M., Heacock, P. & Dowhan, W. (2006) Phosphatidylethanolamine and monoglucosyldiacylglycerol are interchangeable in supporting topogenesis and function of the polytopic membrane protein lactose permease. *J. Biol. Chem.* **281**, 19172–19178
35. Zhang, W., Bogdanov, M., Pi, J., Pittard, A.J. & Dowhan, W. (2003) Reversible topological organization within a polytopic membrane protein is governed by a change in membrane phospholipid composition. *J. Biol. Chem.* **278**, 50128–50135
36. Zhang, W., Campbell, H.A., King, S.C. & Dowhan, W. (2005) Phospholipids as determinants of membrane protein topology: phosphatidylethanolamine is required for the proper topological organization of the γ-aminobutyric acid permease (GabP) of *Escherichia coli*. *J. Biol. Chem.* **280**, 26032–26038
37. Vitrac, H., Bogdanov, M., Heacock, P. and Dowhan, W. (2011) Lipids and topological rules of membrane protein assembly: balance between long- and short-range lipid–protein interactions. J. Biol. Chem., doi:10.1074/jbc.M110.214387
38. von Heijne, G. (2006) Membrane-protein topology. *Nat. Rev. Mol. Cell Biol.* **7**, 909–918
39. Higy, M., Junne, T. & Spiess, M. (2004) Topogenesis of membrane proteins at the endoplasmic reticulum. *Biochemistry* **43**, 12716–12722
40. Gbaguidi, B., Hakizimana, P., Vandenbussche, G. & Ruysschaert, J.M. (2007) Conformational changes in a bacterial multidrug transporter are phosphatidylethanolamine-dependent. *Cell. Mol. Life Sci.* **64**, 1571–1582
41. Andersson, H. & von Heijne, G. (1994) Membrane protein topology: effects of $\Delta\mu$ H$^+$ on the translocation of charged residues explain the 'positive inside' rule. *EMBO J.* **13**, 2267–2272

42. Cao, G., Kuhn, A. & Dalbey, R.E. (1995) The translocation of negatively charged residues across the membrane is driven by the electrochemical potential: evidence for an electrophoresis-like membrane transfer mechanism. *EMBO J.* **14**, 866–875
43. van de Vossenberg, J.L., Albers, S.V., van der Does, C., Driessen, A.J. & van Klompenburg, W. (1998) The positive inside rule is not determined by the polarity of the $\Delta\Psi$ (transmembrane electrical potential). *Mol. Microbiol.* **29**, 1125–1127
44. Seppala, S., Slusky, J.S., Lloris-Garcera, P., Rapp, M. & von Heijne, G. (2010) Control of membrane protein topology by a single C-terminal residue. *Science* **328**, 1698–1700

Biochem. Soc. Symp. 78
Citation reference: Biochem. Soc. Trans. (2011) **39**, 775–779.

9

Mapping lipid and detergent molecules at the surface of membrane proteins

Richard J. Cogdell*[1], Alastair T. Gardiner*, Aleksander W. Roszak†, Sigitas Stončius‡, Pavel Kočovský‡ and Neil W. Isaacs†

**Institute of Molecular Cell and Systems Biology, College of Medical, Veterinary and Life Sciences, Glasgow Biomedical Research Centre, 120 University Place, Glasgow G12 8TA, U.K., †Department of Chemistry, WestChem, University of Glasgow, Glasgow Biomedical Research Centre, 120 University Place, Glasgow G12 8TA, U.K., and ‡Department of Chemistry, WestChem, University of Glasgow, Joseph Black Building, Glasgow G12 8QQ, U.K.*

Abstract

Electron-density maps for the crystal structures of membrane proteins often show features suggesting binding of lipids and/or detergent molecules on the hydrophobic surface, but usually it is difficult to identify the bound molecules. In our studies, heavy-atom-labelled phospholipids and detergents have been used to unequivocally identify these binding sites at the surfaces of test membrane proteins, the reaction centres from *Rhodobacter sphaeroides* and *Blastochloris viridis*. The generality of this method is discussed in the present article.

[1]*To whom correspondence should be addressed (email r.cogdell@bio.gla.ac.uk).*

Introduction

Our previous site-directed mutagenesis study on the relationship between the structure and function of the *Rhodobacter sphaeroides* RC (reaction centre) determined the X-ray crystal structure of many such mutants [1]. Careful analysis of these structures revealed two features in the non-protein electron density at the surface of this membrane protein complex. First, there was a tightly bound region of density that could be successfully fitted with the lipid cardiolipin (diphosphatidylglycerol) [2]. Secondly, there were many regions of undefined electron density, some of which were found to lie parallel to each other and could be fitted with either bound detergent molecules or parts of bound phospholipids. While thinking about how to differentiate between these two possibilities, one of us (R.J.C.) remembered some work by Tony Lee that had used brominated phospholipids produced by the addition of bromine atoms across unsaturated double bonds in the fatty acid tails [3]. The element bromine contains considerably more electrons than the other elements that constitute proteins and lipids and will therefore scatter X-rays more strongly, i.e. bromine is a 'heavy atom'. It was reasoned that, if bromine was added to either phospholipids or detergents, then these atoms should be easily visible in the resulting electron density maps. In this way, it should be possible to distinguish between phospholipids and detergents and be able, hopefully, to characterize their binding sites at the surface of membrane proteins.

Localization of phospholipid-binding sites on the surface of RCs

Figure 1(A) shows the non-protein electron density at a region of the *Rb. sphaeroides* RC surface and the typical appearance of undefined tube-like density can be clearly seen [4]. The *Rb. sphaeroides* RC pigment–protein complex was solubilized, purified and crystallized in the presence of the detergent LDAO (*N*,*N*-dimethyldodecylamine-*N*-oxide). In Figure 1(A), individual tubes of density have each been fitted with one molecule of LDAO; however, they could equally have been fitted with the fatty acyl chains of a phospholipid. In such a case, it is assumed that the headgroup portion of the phosphatidyl molecule is disordered and so is not visible in the electron-density map. Figure 1(B) shows exactly the same region on the surface of the protein (similar, although not identical, view) when the *Rb. sphaeroides* RC was co-crystallized in the presence of dibromo-PC (phosphatidylcholine) [4]. Instead of the two adjacent LDAO molecules, as in Figure 1(A), there is one molecule of dibromo-PC [4]. It is important at this point to emphasize a key point about such X-ray crystal structures. These structures are solved assuming that each RC in the crystal is structurally identical. For the protein portion of the RC, this is probably a safe assumption. However, for the non-protein electron density at the surface of the RC, this is undoubtedly not a good assumption. Rather, some putative

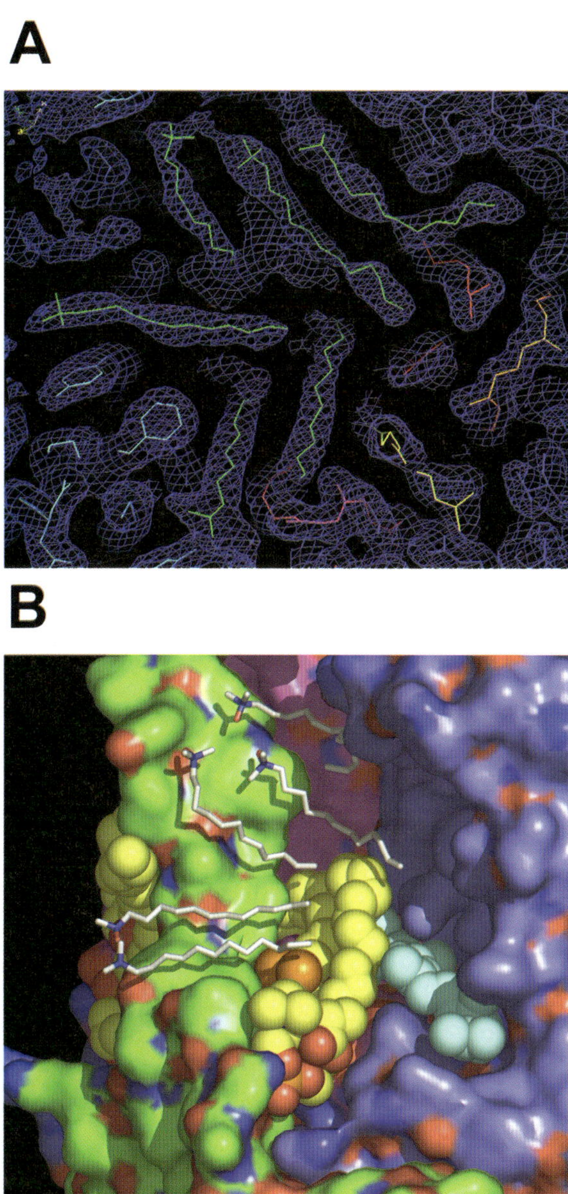

Figure 1. *Rb. sphaeroides* RC surface features within a groove of the transmembrane domain

(**A**) LDAO molecules (green) fitted into surface $2F_o - F_c$ electron density shown at the 0.5σ level at Site A for the 1.95 Å (1 Å = 0.1 nm) structure of the HC(266M) RC mutant from *Rb. sphaeroides* [5]. The whole displayed area is within the transmembrane region of the RC; ubiquinone UQ_A molecule is coloured yellow. (**B**) Site A with bound brominated phospholipid in the groove to the right of (green) helix H of the RC from *Rb. sphaeroides* co-crystallized with the dibromo-PC [4]. The brominated PC is represented as van der Waals spheres with carbon atoms in yellow, oxygen atoms in red, and bromine and phosphorus atoms in orange. Ubiquinone molecule UQ_A is drawn as cyan spheres, and LDAO molecules are shown as white stick-style models. The RC protein

continued on page 112

lipid-binding sites on one RC will be occupied by lipids and some by detergents and, moreover, this will be different on the surface of different RCs in the same crystal. Therefore, when trying to identify putative lipid-binding sites on the surface of membrane proteins, it must be borne in mind that often only partial regions of any given lipid or detergent molecule are sufficiently well ordered to be visible in the electron-density map and also that the final structure at any site will be an average of the different molecules present (or possibly without any molecules present) at that site in the different RCs in the single crystal. There are at least two factors that determine whether a site on the surface of the membrane protein is filled with lipid or detergent or remains empty. The first is the relative 'strength' of the detergent itself. Of course, not all proteins are stable in all detergents, rather some, notably overexpressed mammalian membrane proteins, are only stable in very mild alkyl sugar detergents such as DDM (dodecyl maltoside), whereas more robust membrane proteins, for example some bacterial photosynthetic complexes, can be solubilized, purified and crystallized in much more robust detergents such as LDAO. Thus a delicate balance must be achieved with the detergent so that it disrupts completely the lipid–lipid interactions in the membrane, but does not perturb the interactions within a protein that keep it stable. However, some of the annular lipids may have stronger interactions with the surface of the protein than with the detergent micelle and so will remain in their sites.

The second factor that determines the occupancy of sites on the membrane surface is the conditions chosen for solubilization of the membranes. Generally, a higher concentration of detergent or a higher temperature for solubilization will lead to harsher conditions and the removal of more tightly bound lipids. Again, a balance must be achieved between solubilization conditions that are too harsh and risk denaturing or inactivating the protein in question and those that are too mild, resulting in an insufficiently solubilized protein preparation that contains too many lipids. Such a sample may be too inhomogeneous to be able to form crystals. The concept of binding sites on the surface of the protein with various degrees of affinity for lipids and detergents has consequences for the common procedure of exchanging one detergent for another before crystallization. If the detergent used for solubilization has a high affinity for sites on the protein surface, then it may not be completely removed during detergent exchange, even if the procedure is performed with great diligence.

The occupancy of the site shown in Figure 1(B), designated Site A, with dibromo-PC is approximately 25% and so illustrates the need for heavy-atom tagging. Without the increased X-ray scattering from the bromine atoms and corresponding bulges on the electron-density maps, it would not have

surfaces (including associated pigments) are shown in blue for chain L, green for chain H and pink for chain M. The molecule of cardiolipin is represented in the same colour scheme as dibromo-PC is visible on the left-hand side of (green) helix H. Reprinted with permission from [4]. © 2007 American Chemical Society.

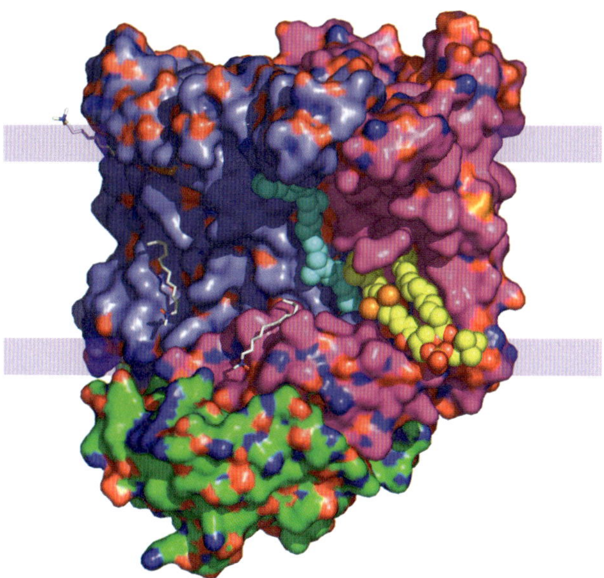

Figure 2. The second lipid-binding site on the surface of the *Rb. sphaeroides* RC
Site B with bound dibromo-PC represented with the van der Waals spheres as in Figure 1(B) at the other side of the RC surface to Site A. Ubiquinone molecule UQ_B is drawn as cyan spheres. The colour schemes for all the remaining structure components are as in Figure 1(B). Reprinted with permission from [4]. © 2007 American Chemical Society.

been possible to definitively assign the density to dibromo-PC. A second site, designated Site B, on the other side of the RC was also found with the dibromo-PC bound (Figure 2).

In order to try to investigate the specificity of lipid-binding sites, such as Site A and Site B, for other phospholipids, the co-crystallization experiment was repeated using phospholipids with different headgroups, such as PG (phosphatidylglycerol), PE (phosphatidylethanolamine) and PS (phosphatidylserine), and with two different positions of the bromine atoms on the fatty acid tails (6,7- or 9,10-dibromination) [4]. All of these lipids were found to bind Site B, whereas their binding in Site A was non-existent or very limited. The occupancy of the brominated phospholipid in Site B was again low (~12–25%), but it was clear that different sites show different degrees of specificity. In this case, Site A would only accept 6,7-dibromo-PC, whereas Site B showed a lower degree of specificity. Comparison of these two sites suggests that Site A has a limited length of the groove on the RC surface, including specific locations of the hydrogen-bond donor groups, and is therefore very specific in the lipid it accepts, whereas Site B is located in the flexible region of the RC (in the extended groove near the site of the distal ubiquinone, UQ_B) and can accept a variety of lipids.

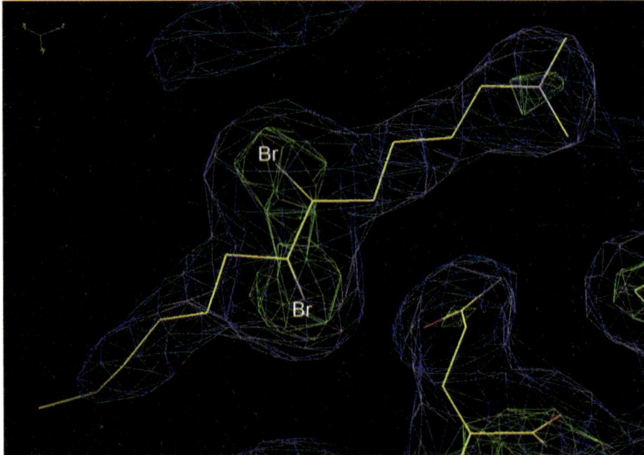

Figure 3. The Br-LDAO molecule in Site A shown in a stick model representation with bromine atoms labelled
To show the extremely high scattering of bromine atoms, the $2F_o - F_c$ electron density is drawn at two levels: the blue contour is at 1.0σ level and the green contour at 4.0σ level.

Binding of heavy-atom derivatives of detergents on the surface of RC from *Rb. sphaeroides* and *Blastochloris viridis*

Br-LDAO is not commercially available and so, in order to carry out the converse experiment, i.e. labelling the detergent rather than the phospholipid, Br-LDAO was synthesized. Crystals of the RC were soaked in Br-LDAO and bromine atoms were looked for in the crystal structure. Figure 3 shows a site on the RC from *Rb. sphaeroides* where a molecule of Br-LDAO is clearly visible in the electron-density map (structural data to be published). This detergent-binding site corresponds to Site A described above, where it was shown that dibromo-PC could bind. This result illustrates the usefulness of heavy-atom-tagged detergents in combination with X-ray crystallography to locate detergent-binding sites on the surface of membrane proteins. Experiments have also recently been started to identify, compare and contrast the binding sites on the surface of the *Blc. viridis* RC with those identified previously on the surface of the *Rb. sphaeroides* RC. An example is given in Figure 4 of initial electron density from a *Blc. viridis* RC and Br-LDAO co-crystallization experiment, fitted with an LDAO molecule usually found at this site. Features of this electron density clearly indicate dibromination in the middle of the aliphatic chain (structural data to be published).

In order to extend this concept, another set of detergents were synthesized incorporating mercury atoms as the heavy atoms rather than bromine. This change was undertaken because it has been suggested that carbon–bromine bonds are susceptible to radiation damage. This damage could possibly contribute to the low occupancy of the bound brominated lipids as the occupancy of the aliphatic atoms in the fatty acid tails was higher than occupancy of the bromine atoms. Figure 5 shows the result of a preliminary experiment

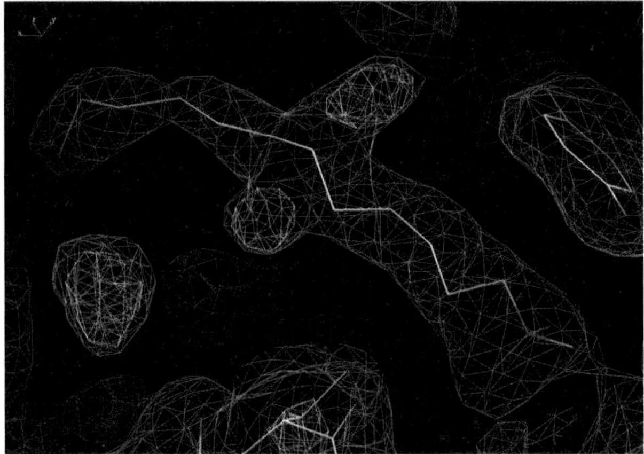

Figure 4. The Br-LDAO-binding site at the surface of RC from *Blc. viridis*
The $2F_o - F_c$ electron density at 1.0σ level (blue) and $F_o - F_c$ electron density at 3.0σ level (green) as found in the initial maps calculated for the 1.85 Å (1 Å = 0.1 nm) structure of *Blc. viridis* RC co-crystallized with the Br-LDAO (yellow model in density is a non-brominated LDAO molecule, which was usually found in this site).

where RCs were co-crystallized with Hg-LDAO. Again the enhanced X-ray scattering of the heavy atom, in this case mercury, is clearly seen in the initial electron-density maps, although the density for the aliphatic atoms of LDAO are not visible next to the very high mercury-density area (structural data to be published). Further work on these mercury-tagged detergents is continuing.

Investigating an idea of phasing of X-ray diffraction data for membrane proteins through the use of heavy-atom derivatives of lipids and detergents

In the study described, the usefulness of the heavy-atom detergent derivatives mentioned was tested for phasing of the X-ray diffraction data. The aim was to establish whether these detergent derivatives could be used as 'easy/natural' heavy-atom compounds for phasing of diffraction data for membrane protein crystals using anomalous dispersion methods. Membrane proteins require detergents for solubilization and stability, so replacement of these detergents by heavy-atom-containing derivatives would be a much simpler and straightforward method to introduce the heavy atom into the crystal than the toxic and hazardous heavy-atom compound soaks required at present. Additionally, the RC electron-density features suggest that detergent molecules bind in the grooves of the membrane protein's hydrophobic surface (Figure 1A). However, the limited degree of occupancy compounded by possible effects resulting from X-ray damage to the carbon–bromine bond mean that the anomalous signal produced by these heavy atoms is not yet sufficient to allow successful phasing. Further efforts are being undertaken to (i) increase the level

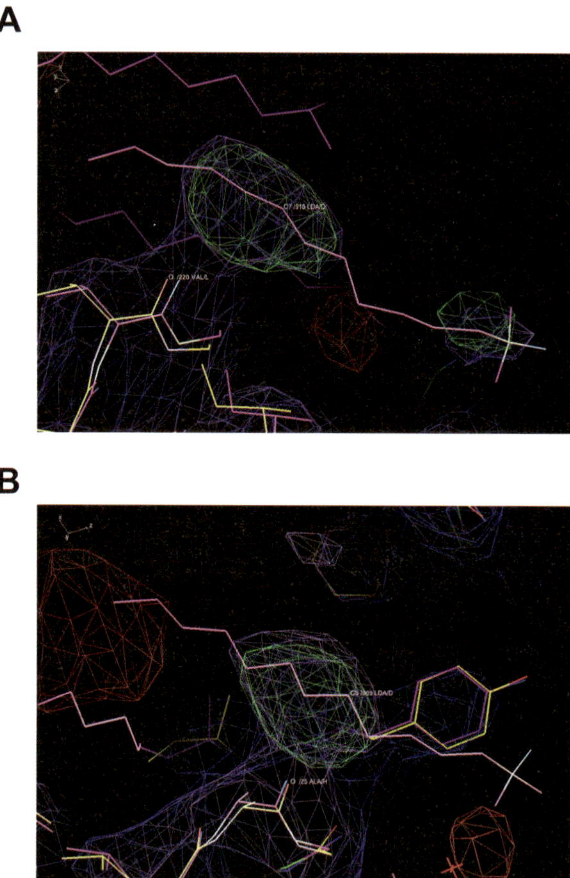

Figure 5. Binding of Hg-LDAO molecules at the surface of RC from *Rb. sphaeroides*
The $2F_o - F_c$ electron density at 1.0σ level (blue) and $F_o - F_c$ electron density at 3.0σ level (green) as found in the initial maps calculated for the 3.25 Å (1 Å = 0.1 nm) structure of RC co-crystallized with Hg-LDAO (model in yellow). Also shown (in magenta) is the superimposed 1.95 Å structure of the HC(266M) RC mutant from *Rb. sphaeroides* [5] with the LDAO molecule passing through the high blue/green unmodelled density. (**A**) Electron density visible in Site B for a putative mercury atom of the Hg-LDAO molecule in a position to co-ordinate the carbonyl oxygen of the valine residue at position 220 of the L subunit (L220-Val). (**B**) Electron density for a putative mercury atom of the Hg-LDAO molecule residing in the wide groove of which Site A is part. This density coincides with the LDAO molecule of the HC(266M) RC mutant, shown in Figure 1(A) as the most horizontally oriented of the modelled (green) LDAO molecules. The mercury atom in this position is co-ordinating carbonyl oxygen of the alanine residue at position 25 of the H subunit (H25-Ala).

of occupancy, and (ii) modify the data-collection strategy in order to reduce the experimental errors, which should increase the chances of successful phasing.

Discussion

The idea behind the study described was inspired by previous work carried out by Tony Lee. It is a great pleasure therefore to dedicate this short contribution to him on the occasion of his retirement. This study has shown the usefulness of heavy-atom-tagged phospholipids and detergents in combination with X-ray crystallography to explore surface binding sites for these molecules around membrane proteins. Of course, the molecules in these binding sites must be sufficiently well ordered to be visualized in the electron-density maps. Often, only portions of these molecules can be seen in the crystal structures and then identification becomes very difficult. The heavy-atom tagging, when successful, allows unequivocal identification even when only part of the molecule can be seen. These molecules are a very useful addition to a crystallographer's toolkit.

Funding

This research was supported by the Biotechnology and Biological Sciences Research Council both as part of the Membrane Protein Structure Initiative (MPS*i*) and grant number BB/E008720/1 as well as the Wellcome Trust. We acknowledge the European Synchrotron Radiation Facility (ESRF), Grenoble, France, and the Diamond Light Source, Oxfordshire, U.K., for beamtime.

References

1. McAuley-Hecht, K.E., Fyfe, P.K., Ridge, J.P., Prince, S.M., Hunter, C.N., Isaacs, N.W., Cogdell, R.J. & Jones, M.R. (1998) Structural studies of wild-type and mutant reaction centers from an antenna-deficient strain of *Rhodobacter sphaeroides*: monitoring the optical properties of the complex from bacterial cell to crystal. *Biochemistry* **37**, 4740–4750
2. McAuley, K.E., Fyfe, P.K., Ridge, J.P., Isaacs, N.W., Cogdell, R.J. & Jones, M.R. (1999) Structural details of an interaction between cardiolipin and an integral membrane protein. *Proc. Natl. Acad. Sci. U.S.A.* **96**, 14706–14711
3. East, J.M. & Lee, A.G. (1982) Lipid selectivity of the magnesium-ion dependent adenosine-triphosphate, studied with fluorescence quenching by a brominated phospholipid. *Biochemistry* **21**, 4144–4151
4. Roszak, A.W., Gardiner, A.T., Isaacs, N.W. & Cogdell, R.J. (2007) Brominated lipids identify lipid binding sites on the surface of reaction center from *Rhodobacter sphaeroides*. *Biochemistry* **46**, 2909–2916
5. Gardiner, A.T., Roszak, A.W. & Cogdell, R.J. (2005) The structure of the ZnRC HC(266M) at RT and 100K., 1st Joint German/British Bioenergetics Conference 'Mechanisms of Bioenergetic Membrane Proteins: Structures and Beyond', 20–24 March 2005, Wiesbaden, Germany (Abstract)

Biochem. Soc. Symp. 78
Citation reference: Biochem. Soc. Trans. (2011) **39**, 781–787.

10

Improving cardiac Ca^{2+} transport into the sarcoplasmic reticulum in heart failure: lessons from the ubiquitous SERCA2b Ca^{2+} pump

Peter Vangheluwe[1] and Frank Wuytack
Laboratory of Cellular Transport Systems, Department of Molecular Cell Biology, K.U. Leuven, Herestraat 49, bus 802, B-3000 Leuven, Belgium

Abstract

As a major Ca^{2+} pump in the sarcoplasmic reticulum of the cardiomyocyte, SERCA2a (sarcoplasmic/endoplasmic reticulum Ca^{2+}-ATPase 2a) controls the relaxation and contraction of the cardiomyocyte. It is meticulously regulated by adapting its expression levels and affinity for Ca^{2+} ions to the physiological demand of the heart. Dysregulation of the SERCA2a activity entails poor cardiomyocyte contractility, resulting in heart failure. Conversely, improving cardiac SERCA2a activity, e.g. by boosting its expression level or by increasing its affinity for Ca^{2+}, is a promising strategy to rescue contractile dysfunction of the failing heart. The structures of the related SERCA1a Ca^{2+} pump and the Na^+/K^+-ATPase of the plasma membrane exposed the pumping mechanism and conserved domain architecture of these ion pumps. However, how the Ca^{2+}

[1]*To whom correspondence should be addressed (email peter.vangheluwe@med.kuleuven.be).*

affinity of SERCA2a is regulated at the molecular level remained unclear. A structural and functional analysis of the closely related SERCA2b Ca^{2+} pump, i.e. the housekeeping Ca^{2+} pump found in the endoplasmic reticulum and the only SERCA isoform characterized by a high Ca^{2+} affinity, aimed to fill this gap. We demonstrated the existence of a novel and highly conserved site on the SERCA2 pump mediating Ca^{2+} affinity regulation by the unique C-terminus of SERCA2b (2b-tail). It differs from the earlier-described target site of the affinity regulator phospholamban. Targeting this novel site may provide a new approach to improve SERCA2a function in the failing heart. Strikingly, the intramembrane interaction site of the 2b-tail in SERCA2b shares sequence and structural homology with the binding site of the β-subunit on the α Na^+/K^+-ATPase. Thus P-type ATPases seem to have developed related mechanisms of regulation, and it is a future challenge for us to discover these general principles of P-type regulation.

Regulation of SERCA2a activity controls cardiac contractility

Muscle contraction is initiated by a transient release of Ca^{2+} ions from the SR (sarcoplasmic reticulum) into the cytosol [1]. The Ca^{2+} content of the SR will determine how much Ca^{2+} can be released for contraction, which makes it an important determinant of contractile strength. In cardiac muscle, Ca^{2+} uptake into the SR is almost exclusively catalysed by SERCA2a (sarcoplasmic/endoplasmic reticulum Ca^{2+}-ATPase 2a), which couples the energy of ATP hydrolysis to the transport of two Ca^{2+} ions across the SR membrane. SERCA2a not only determines the amount of Ca^{2+} in the SR available to activate the next contraction, but also controls to a large extent the rate of cardiac relaxation. After a Ca^{2+} release event in mammalian ventricular myocytes, SERCA2a transports approximately 50–90 % of the Ca^{2+} back into the SR [1].

The expression level of SERCA2a significantly affects cardiac SR Ca^{2+} uptake and overall cardiac contractile properties [2]. But it is the acute regulation of SERCA2a activity that allows a fast beat-to-beat control of the contractile properties of the heart. It is therefore not surprising that the activity of the Ca^{2+} pump is under strict control and is constantly adjusted to meet the physiological demand of the heart [3,4].

SERCA2 is encoded by the *ATP2A2* gene. First, the level of expression and alternative splicing are controlled at the transcript level [4]. In general, there is a strong correlation between the SERCA2a mRNA and protein levels and the number of transcripts will determine the protein level and overall capacity of Ca^{2+} transport [2]. Both the promoter activity [5] and mRNA stability [6] are tightly regulated in the heart. However, adjusting the expression level to the physiological demand remains a very slow process in view of the slow turnover of the SERCA2a protein (an estimated 2–3 day half-life time) [7]. Besides its

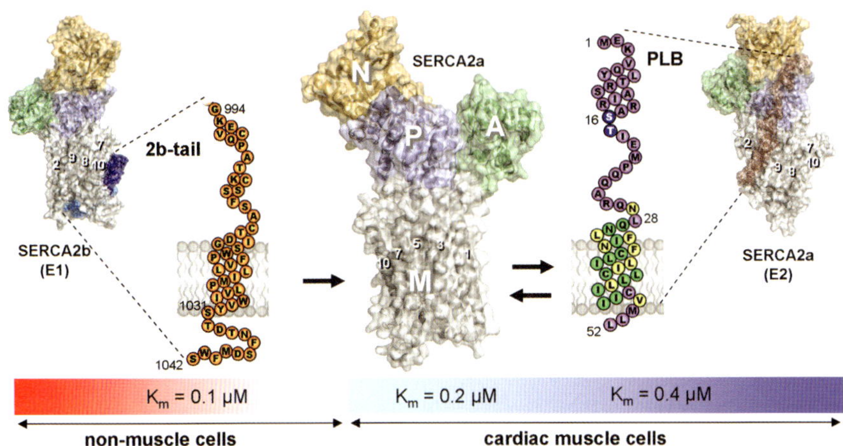

Figure 1. Regulation of SERCA2a's Ca^{2+} affinity strongly affects cardiomyocyte contractility

Early in cardiac development, the housekeeping variant SERCA2b is replaced by the muscle-specific variant SERCA2a via alternative splicing of the SERCA2 mRNA. Compared with SERCA2a, SERCA2b contains a 49-residue-long C-terminus (2b-tail, dark blue on the left-hand structure) that increases the Ca^{2+} affinity and reduces the maximal turnover rate of the pump. In the cardiomyocyte, PLB (red on the right-hand structure) is expressed which dynamically regulates the Ca^{2+} affinity of the pump. Direct interaction of PLB inhibits the pump by reducing the Ca^{2+} affinity, whereas phosphorylation of PLB during β-adrenergic stimulation relieves the inhibition. The binding sites of the 2b-tail and PLB are indicated by numbering the transmembrane segments on the insets of the SERCA2 pump. The apparent Ca^{2+} affinity is represented by the K_m value of Ca^{2+}-dependent Ca^{2+}-transport. A high K_m value of the pump indicates a low apparent affinity for Ca^{2+} ions. Light orange, nucleotide-binding domain (N); blue, phosphorylation domain (P); green, actuator domain (A); grey, transmembrane domain (M).

expression level, alternative processing (splicing and polyadenylation) of the *ATP2A2* transcript is also strictly controlled. Early in the developing heart, the *ATP2A2* transcript is alternatively processed by a so far poorly understood, but highly conserved, mechanism [4,8]. This results in the expression of the more specialized and muscle-specific SERCA2a variant instead of the housekeeping SERCA2b variant that is present in all cells (Figure 1). In SERCA2a, the 49-amino-acid-long C-terminus of SERCA2b (2b-tail) is replaced by a much shorter, four-amino-acid, tail. Without this 2b-tail, SERCA2a displays a 2-fold higher maximal turnover rate, but a lower apparent Ca^{2+} affinity compared with SERCA2b [9]. The physiological importance of this cardiac isoform switch was investigated in-depth in a genetically modified mouse model in which the alternative *ATP2A2* processing towards the SERCA2a isoform was prevented. This resulted in a replacement of SERCA2a by SERCA2b in the heart that was accompanied by a somewhat unexpected down-regulation of total cardiac SERCA2 levels by approximately 40 % [10]. Although these mice developed mild concentric left ventricular hypertrophy, the SERCA2b Ca^{2+} pump is able to preserve normal contractile function despite its lower expression levels [10–12]. This questioned why the heart expresses the specialized SERCA2a isoform in the first place. Recent simulations of Ca^{2+} signalling in a murine cardiomyocyte now suggest that, under resting conditions, i.e. at a low adrenergic tone, the SERCA2a pump performs less efficiently than the higher-Ca^{2+}-affinity variant SERCA2b owing to a higher pumping activity of SERCA2b at low cytosolic Ca^{2+} concentrations (50–200 nM) [13]. However, SERCA2a would represent a better compromise between performance in resting conditions and during β-adrenergic stress when the cytosolic Ca^{2+} load significantly increases. Indeed, Ca^{2+} removal by the SERCA2a pump is more efficient at high cytosolic Ca^{2+} load due to its higher maximal pumping rate and lower apparent affinity for Ca^{2+} [13]. Experimental evidence supports this hypothesis, since SR Ca^{2+} uptake in the SERCA2b mice is impaired at high Ca^{2+} loads and during β-adrenergic stimulation [11,12]. Thus SERCA2a appears to be specifically adapted to work under a broader dynamic range of cytosolic Ca^{2+} concentrations from rest to maximal adrenergic stimulation [13].

A second, more rapid, mechanism of SR Ca^{2+} uptake control is the dynamic regulation of the pump's apparent affinity for Ca^{2+} ions [14]. Although the cardiac SERCA2a isoform already operates with a lower Ca^{2+} affinity than the SERCA2b variant found in most other cells, SERCA2a's apparent Ca^{2+} affinity can be reduced even further by the direct interaction with one or both of two related small TM (transmembrane) proteins, i.e. PLB (phospholamban) in the atria and ventricles and sarcolipin in the atria [15] (Figure 1). PLB and sarcolipin act as inhibitors of the SR Ca^{2+}-ATPase, but this inhibition is relieved during β-adrenergic stress by phosphorylation of PLB on Ser^{16} and Thr^{17} [16], or of sarcolipin presumably on Thr^{5} [17]. This results in a very rapid and robust increase in SR Ca^{2+} uptake, with a strong impact on cardiac relaxation and contraction, making PLB and, presumably, also sarcolipin important mediators of the cardiac stress response [15]. PLB and/or sarcolipin fit into a hydrophobic cleft found in the TM region of the Ca^{2+} pump that is formed by TM helices 2, 4, 6 and 9 [18,19]. By counteracting the closure of the hydrophobic cleft that normally follows the binding of Ca^{2+} to the pump, this interaction forces the pump to dwell longer in its Ca^{2+} free conformation (E2). Thus these inhibitors exert a kinetic effect on the pump's catalytic cycle that decreases the apparent affinity for cytosolic Ca^{2+} [15].

Finally, an increasing number of other protein interactors (such as HAX-1, HRC or S100A1) and post-translational modifications (such as nitrosylation, phosphorylation or SUMOylation) have now been discovered, which alter the pump's stability and activity, further highlighting the central role of SERCA2a in cardiac function (reviewed in depth elsewhere [3,4]).

Abnormal SERCA2a activity in heart failure

Heart failure is the leading cause of hospitalization and death in the developed world and is defined as the inability of the heart to supply sufficient blood flow to meet the body's needs. The poor contractile properties of the failing cardiomyocytes are typically associated with abnormal Ca^{2+} cycling such as a reduced Ca^{2+} content in the SR [15,20,21]. In part, this is reflected by impaired SR Ca^{2+} uptake because of a lower SERCA2a expression level. Together with unchanged PLB expression levels, the PLB/SERCA2a ratio increases in conditions of heart failure, leading to a lower apparent Ca^{2+} affinity of the pump at rest. Moreover, the lower degree of PLB phosphorylation was also reported as a possible complicating factor contributing to the impaired cardiac contractility of the failing heart [4,15,21].

The importance of a proper control of the Ca^{2+} affinity is underscored further by the discovery of several PLB mutants triggering the onset of familial dilated cardiomyopathy. PLB mutations causing a chronic reduction in the apparent Ca^{2+} affinity lead to depressed SR Ca^{2+} uptake and heart failure [22–24]. Moreover, and quite unexpectedly, a higher apparent Ca^{2+} affinity resulting from the absence of PLB also triggers heart failure in humans [25], although not in mice [26]. In mice, cardiomyopathy is only observed upon a combined deletion of PLB and sarcolipin [27] or expression of the high-Ca^{2+}-affinity variant SERCA2b [10,12]. Both mouse models develop cardiac hypertrophy. These studies show that the apparent Ca^{2+} affinity of SERCA2a in cardiac SR needs to be controlled within a tight window to ensure normal cardiac function and growth [4,21].

In the last decade, several strategies were explored in cellular and animal models to improve the contractile properties of the cardiomyocyte by increasing the activity of SERCA2a [20]. One promising approach in humans is to increase SERCA2a expression via adeno-associated viral gene transfer of SERCA2a under control of a cardiac-specific myosin heavy chain α promoter (in Phase II clinical trial, CUPID study) [28,29]. The observation that, in the normal physiological setting, SERCA2a activity in the heart is dynamically adjusted to the physiological demand by modulating its affinity for Ca^{2+} suggests that other interventions, aimed to increase SERCA2a's Ca^{2+} affinity, represent another strategy to effectively improve the activity of the pump [15,30]. The proof-of-concept was provided in several animal models in which the inhibition of the pump by PLB was diminished, e.g. by long-term cardiac-targeted RNA interference of PLB [30] or by overexpression of pseudo-phosphorylated constitutively active PLB (PLB-S16E) [31].

Structure–function relationship of the ubiquitous SERCA2b pump

Since controlling the apparent Ca^{2+} affinity of the Ca^{2+} pump is of paramount importance for cardiac function, we elaborate on the molecular mechanisms of Ca^{2+}-affinity regulation.

Already more than 10 years ago, the first high-resolution crystal structure of the fast-twitch skeletal-muscle isoform SERCA1a was published [32]. More

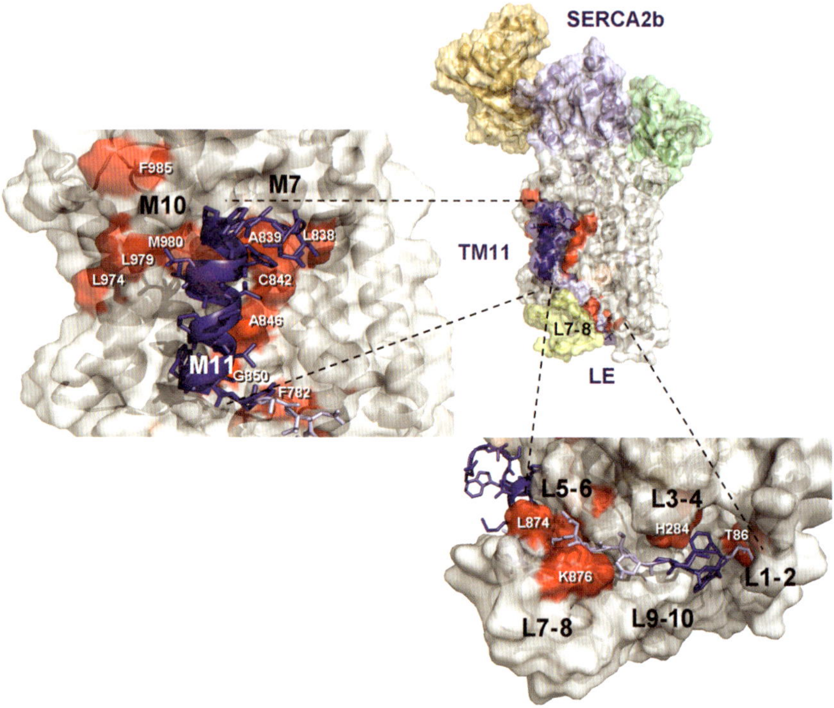

Figure 2. Detailed view of the 2b-tail-binding site on the SERCA2b structural model
Top right: structural model of SERCA2b in the Ca^{2+}-bound E1 conformation [39]. Luminal loop L7–L8 is indicated in light yellow, and is a crucial interaction site of the 2b-tail. Light orange, N-domain; blue, P-domain; green, A-domain; grey, M-domain. Left: detailed view of the interaction site of the extra TM segment TM11 (dark blue) on the region TM7 and TM10 of the SERCA pump. Bottom right panel: detailed view of the luminal interaction site of the 2b-tail. Dark blue residues close to luminal loops L1–L2 and L3–L4 correspond to the residues M^{1039}FWS of the luminal 2b-tail. In red, the residues that are important for 2b-tail function; mutating these residues reduces the apparent affinity of SERCA2b, but not that of SERCA2a.

recently, structures of other archetypal P-type ATPases were solved (Na^+/K^+-ATPase [33,34] and H^+-ATPase [35]), indicating that the general architecture of the P-type ATPases is very well conserved and contains four typical domains: a transmembrane domain (M) containing the binding sites of the transported ions, and three clearly defined cytosolic domains [a nucleotide-binding (N), phosphorylation (P) and actuator (A) domain] (Figure 1).

So far, over 20 high-resolution crystal structures of SERCA1a have been published covering almost all conformations of the kinetic cycle, yielding detailed molecular insights into the Ca^{2+}-pumping mechanism (reviewed in [36–38]). ATP binding occurs on the N-domain, whereas the P-domain catalyses ATP hydrolysis. The P-domain contains a key aspartate residue, which is the acceptor of the γ-phosphate of ATP leading to autophosphorylation of the pump when Ca^{2+} is bound (hence the name P-type ATPase). Later in the cycle, the phosphorylated aspartate residue becomes dephosphorylated by the A-domain via a conserved glutamate residue, which catalyses the event. The phosphorylated aspartate residue is temporarily protected against dephosphorylation, but, since the A-domain rotates in the pump cycle, the catalysing glutamate residue becomes properly positioned to catalyse the dephosphorylation. Importantly, the phosphorylation and dephosphorylation of the pump is tightly linked with the closure of the cytosolic or luminal gate respectively, restricting access to the Ca^{2+} binding site. This coupling ensures that both gates are never open at the same time, i.e. first, the cytosolic gate closes and Ca^{2+} ions are excluded before opening of the luminal gate can occur. This sequential opening is crucial to generate a 10000-fold Ca^{2+} gradient across the SR membrane [36–38].

Less structural information is available for other SERCA isoforms. The SERCA2 proteins, with the variants SERCA2a and SERCA2b, share over 80 % sequence identity with SERCA1a, which points to a common Ca^{2+}-pumping mechanism [39]. But the SERCA2b isoform differs from SERCA1a or SERCA2a by three related structural and functional features: SERCA2b has an extended 2b-tail and it presents a 2-fold higher affinity for cytosolic Ca^{2+} ions and only half the maximal turnover rate of the two other isoforms. The 2b-tail comprises an additional TM segment (TM11), so that SERCA2b is the only known Ca^{2+} pump with 11 TM helices. The 2b-tail ends with a small luminal extension comprising 12 residues S^{1031}TDTNFSDMFWS^{1042} [9,39–41].

Functional measurements on SERCA2b mutants (e.g. alanine substitutions and partial truncations of the 2b-tail) revealed that both TM11 and the luminal extension contribute to the functional effect of the 2b-tail [39] (Figure 2). First, mutations of any of the four C-terminal residues M^{1039}FWS of the luminal extension to alanine affect the Ca^{2+} affinity of SERCA2b. Secondly, the functional effect of the 2b-tail is partially preserved when 11 residues of the luminal extension are removed (T1032*), indicating that not only MFWS, but also a second region upstream of the luminal extension is alsoimportant (presumably TM11) [39]. In contrast, when all 12 residues of the luminal extension are truncated (S1031* mutation) the effect of the 2b-tail is completely abolished [39,40]. Since the S1031A substitution has no functional effect, we hypothesize that TM11 is a second functional region that loses its function or membrane insertion when it is truncated at Ser^{1031}, i.e. too close to the membrane [39].

When the 2b-tail is fused to the SERCA1a isoform (forming a SERCA1a2b chimaera), it partially increased the apparent Ca^{2+} affinity of SERCA1a [39]. Further experiments pointed out that TM11, but not the luminal extension, is responsible for this effect, showing that the site where TM11 interacts with more upstream elements of the pump is conserved between SERCA1a and SERCA2. On the other hand, the luminal binding site for interaction with the luminal extension is not well conserved between both isoforms. The binding site for the luminal extension can only be established in SERCA1a2b provided that the luminal loop of L7–L8 of SERCA2b is inserted to replace the corresponding

SERCA1a loop. Indeed, this substitution leads to a gain-of-function of the 2b-tail in the SERCA1a2b chimaera, providing strong evidence that L7–L8 is an important interaction site for the luminal extension. We subsequently found two substitutions in L7–L8 of SERCA1a (L873M and K876T) that are responsible for the inability of the SERCA1a L7–L8 to interact with the luminal extension of the 2b-tail [39] (Figure 2).

Apparently, TM11 does not disturb the interactions or movements of other TM helices that are required for normal pump function. Therefore the interaction of TM11 probably takes place on a conserved site that is exposed to the lipid environment in the TM domain of the pump. Mutations of exposed and conserved residues were tested in several TM helices, but only alanine substitutions of residues in TM7 and TM10 led to a reduced Ca^{2+} affinity [39] (Figure 2). On the basis of the published SERCA1a crystal structures [32] and our own solved NMR structure of TM11 [39], which is a short α-helix, a structural model for SERCA2b was proposed that is in agreement with the mutagenesis results (Figures 1 and 2). According to that model, TM11 interacts with a region formed by TM7 and TM10 of the pump, opening at its luminal side a groove between loops L5–L6 and L7–L8. This groove is important for the descent of the luminal extension towards a luminal pocket formed by all five luminal loops where the binding of residues M^{1039}FWS takes place. The model provides an elegant explanation for the crucial effect of Leu^{873} and Lys^{876} on L7–L8 of SERCA2b, which are also found in this groove close to the luminal extension [39].

According to some early reports, the chaperones calreticulin and calnexin would interact at the luminal side of SERCA2b with the residues N^{1035}FS of the luminal extension of the 2b-tail [42,43]. It was stated that the interaction with calreticulin or calnexin is required to render the specific enzymatic properties of SERCA2b. Since no N-glycosylation was observed on Asn^{1035}, it was suggested that this interaction occurs in a glycosylation-independent manner [43]. Although we cannot rule out that calreticulin or calnexin interacts with the 2b-tail, our results question whether this interaction is crucial to render the high Ca^{2+} affinity of SERCA2b. Indeed, we show that alanine mutants of the N^{1035}FS region retain normal Ca^{2+}-dependent ATPase activity when overexpressed in COS cells [39]. Moreover, according to the SERCA2b molecular model, the 2b-tail is buried in luminal loops of the pump, making its interaction with other proteins less likely [39].

The mechanism by which the 2b-tail would increase the Ca^{2+} affinity of the pump via an intramolecular interaction was explored further. We demonstrated that the luminal extension of the 2b-tail stabilizes the pump in the Ca^{2+}-bound E1 conformation with high-affinity binding sites facing the cytosol. This fits with the proposed binding site of the C-terminal tetrapeptide of the luminal extension, which is situated in a luminal pocket formed by the five luminal loops of the pump. Mathematical modelling confirmed that holding theenzyme longer in the E1 conformation could explain the increased apparent affinity for Ca^{2+}

[39]. Moreover, structures of the SERCA1a pump in different conformations predict extensive rearrangements of the proposed luminal docking site of the luminal extension. Thus the binding of the luminal extension would slow down these rearrangements, which is in agreement with the experimentally observed slower E1-P–E2-P and E2-P–E2 transitions [39,41].

How the short TM11 α-helix alters the enzymatic properties of the pump needs further exploration, since TM11 binds at a remote and relatively immobile part of the TM region far from the catalytically important regions [39]. One striking observation is, however, the resemblance between the binding site of TM11 on the SERCA pump and the TM region of the β-subunit on the α-subunit of Na^+/K^+-ATPase, which is a known regulator of the apparent K^+ affinity of this pump [33,34,44]. Note that TM11 and the β-subunit independently originated in evolution, but apparently evolved to occupy existing hotspots for regulation on the related P-type ATPases SERCA2b and Na^+/K^+-ATPase [39]. Confirming this common mode of action would point to an exciting example of convergent evolution in the P-type ATPases. The same holds true for PLB and phospholemman (a member of the FXYD family also known as the γ-subunit of the Na^+/K^+-ATPase), which, although they originated independently in evolution, also occupy approximately the same binding site close to TM9 with similar modes of action and regulation [33,34]. Thus P-type ATPases not only share a general architecture and domain organization, but also seem to have developed related mechanisms of regulation. It is a future challenge for us to discover these general principles of P-type regulation.

Purification of the ubiquitous SERCA2b pump

Another future goal is to verify the structural model by solving the atomic resolution structure of the SERCA2b pump. Although now readily feasible for SERCA1a, the much lower endogenous tissue expression has so far hampered purification efforts with SERCA2b [45]. We have developed anew affinity-purification strategy based on thapsigargin, a specific and highly potent inhibitor of SERCA (in the sub-nanomolar range) [46,47]. Thapsigargin binds in a hydrophobic funnel-like cavity of the Ca^{2+} pump formed by TM helices 3, 5 and 7 and inhibits the pump by preventing the movement of these helices relative to each other [47,48]. A thapsigargin analogue was synthesized by S.B. Christensen (Copenhagen, Denmark) to allow coupling of the inhibitor to a Sepharose matrix for affinity chromatography of recombinantly expressed SERCA2b pump [45]. A second affinity-purification step, based on ATP analogues (reactive dyes) coupled to a carrier, was subsequently applied for further purification, concentration and detergent exchange. Without the necessity of adding tags that might interfere with pump function and structure, these affinity chromatography techniques allow the selective purification of correctly folded and functionally active SERCA2b. Note that amino acids forming the thapsigargin-binding pocket in SERCA1a are partially conserved in SPCA1 (secretory pathway

Ca^{2+}-ATPase 1) [45]. SPCA1 therefore also shows some sensitivity towards thapsigargin, but in a thapsigargin concentration range at least 1000-fold higher (micromolar concentration) than SERCA [38]. Interestingly, the successive thapsigargin and reactive dye purification could also be successfully employed to purify SPCA1 from the yeast overexpression system. The selective purification of SERCA and SPCA isoforms will enable future crystallization attempts of these housekeeping Ca^{2+} pumps.

Conclusions

The diminished cardiac contractility in heart failure is partially explained by a reduced activity of the cardiac SERCA2a pump. Improving the expression or function of the Ca^{2+} pump are potential strategies to improve cardiac contractility. One possible strategy is to increase the affinity of the pump for Ca^{2+} ions. We have studied the molecular mechanism of the unique C-terminus of the ubiquitous SERCA2b isoform to obtain insights into the mechanisms of affinity regulation. We have developed a structural model of SERCA2b that explains extensive mutagenesis results and the mode of action of the 2b-tail. Finally, we have described the use of thapsigargin, a selective and potent SERCA inhibitor, to purify SERCA2b for subsequent structural studies.

Interestingly, the predicted docking site in SERCA2 for the TM11 region of the 2b-tail strikingly resembles the corresponding interaction site for the β-subunit on the α-subunit of the Na^+/K^+-ATPase, but it is distinct from the PLB site, which is situated at the opposite side of SERCA (Figure 1). We therefore hypothesize that targeting the 2b-tail interaction site may represent a novel approach to improve the activity of SERCA2a in heart failure, independently of PLB.

Funding

This research was financed by the Research Programme of the Flanders Research Foundation (FWO-Vlaanderen) [grant number G.0646.08] and by the Interuniversity Attraction Poles Programme P6/28 of the Belgian State, Federal Office for Scientific Technical and Cultural Affairs. P.V. is a postdoctoral fellow of the Flanders Research Foundation (FWO-Vlaanderen).

References

1. Bers, D.M. (2002) Cardiac excitation–contraction coupling. *Nature* **415**, 198–205
2. Periasamy, M. & Huke, S. (2001) SERCA pump level is a critical determinant of Ca^{2+} homeostasis and cardiac contractility. *J. Mol. Cell. Cardiol.* **33**, 1053–1063
3. Vangheluwe, P., Raeymaekers, L., Dode, L. & Wuytack, F. (2005) Modulating sarco(endo)plasmic reticulum Ca^{2+} ATPase 2 (SERCA2) activity: cell biological implications. *Cell Calcium* **38**, 291–302
4. Vandecaetsbeek, I., Raeymaekers, L., Wuytack, F. & Vangheluwe, P. (2009) Factors controlling the activity of the SERCA2a pump in the normal and failing heart. *Biofactors* **35**, 484–499
5. Zarain-Herzberg, A. (2006) Regulation of the sarcoplasmic reticulum Ca^{2+}-ATPase expression in the hypertrophic and failing heart. *Can. J. Physiol. Pharmacol.* **84**, 509–521

6. Misquitta, C.M., Chen, T. & Grover, A.K. (2006) Control of protein expression through mRNA stability in calcium signalling. *Cell Calcium* **40**, 329–346
7. Andersson, K.B., Birkeland, J.A., Finsen, A.V., Louch, W.E., Sjaastad, I., Wang, Y., Chen, J., Molkentin, J.D., Chien, K.R., Sejersted, O.M. & Christensen, G. (2009) Moderate heart dysfunction in mice with inducible cardiomyocyte-specific excision of the *Serca2* gene. *J. Mol. Cell. Cardiol.* **47**, 180–187
8. Van Den Bosch, L., Mertens, L., Gijsbers, S., Heyen, M.V., Wuytack, F. & Eggermont, J. (1997) Sequence elements surrounding the acceptor site suppress alternative splicing of the sarco/endoplasmic reticulum Ca^{2+}-ATPase 2 gene transcript. *Biochem. J.* **322**, 885–891
9. Lytton, J., Westlin, M., Burk, S.E., Shull, G.E. & MacLennan, D.H. (1992) Functional comparisons between isoforms of the sarcoplasmic or endoplasmic reticulum family of calcium pumps. *J. Biol. Chem.* **267**, 14483–14489
10. Ver Heyen, M., Heymans, S., Antoons, G., Reed, T., Periasamy, M., Awede, B., Lebacq, J., Vangheluwe, P., Dewerchin, M., Collen, D. et al. (2001) Replacement of the muscle-specific sarcoplasmic reticulum Ca^{2+}-ATPase isoform SERCA2a by the nonmuscle SERCA2b homologue causes mild concentric hypertrophy and impairs contraction–relaxation of the heart. *Circ. Res.* **89**, 838–846
11. Antoons, G., Ver Heyen, M., Raeymaekers, L., Vangheluwe, P., Wuytack, F. & Sipido, K.R. (2003) Ca^{2+} uptake by the sarcoplasmic reticulum in ventricular myocytes of the *SERCA2*$^{b/b}$ mouse is impaired at higher Ca^{2+} loads only. *Circ. Res.* **92**, 881–887
12. Vangheluwe, P., Tjwa, M., Van Den Bergh, A., Louch, W.E., Beullens, M., Dode, L., Carmeliet, P., Kranias, E., Herijgers, P., Sipido, K.R. et al. (2006) A SERCA2 pump with an increased Ca^{2+} affinity can lead to severe cardiac hypertrophy, stress intolerance and reduced life span. *J. Mol. Cell. Cardiol.* **41**, 308–317
13. Raeymaekers, L., Vandecaetsbeek, I., Wuytack, F. & Vangheluwe, P. (2011) Modeling Ca^{2+} dynamics of mouse cardiac cells points to a critical role of SERCA's affinity for Ca^{2+}. *Biophys. J.* **100**, 1216–1225
14. Vangheluwe, P., Schuermans, M., Raeymaekers, L. & Wuytack, F. (2007) Tight interplay between the Ca^{2+} affinity of the cardiac SERCA2 Ca^{2+} pump and the SERCA2 expression level. *Cell Calcium* **42**, 281–289
15. MacLennan, D.H. & Kranias, E.G. (2003) Phospholamban: a crucial regulator of cardiac contractility. *Nat. Rev. Mol. Cell Biol.* **4**, 566–577
16. MacLennan, D.H., Asahi, M. & Tupling, A.R. (2003) The regulation of SERCA-type pumps by phospholamban and sarcolipin. *Ann. N.Y. Acad. Sci.* **986**, 472–480
17. Bhupathy, P., Babu, G.J., Ito, M. & Periasamy, M. (2009) Threonine-5 at the N-terminus can modulate sarcolipin function in cardiac myocytes. *J. Mol. Cell. Cardiol.* **47**, 723–729
18. Toyoshima, C., Asahi, M., Sugita, Y., Khanna, R., Tsuda, T. & MacLennan, D.H. (2003) Modeling of the inhibitory interaction of phospholamban with the Ca^{2+} ATPase. *Proc. Natl. Acad. Sci. U.S.A.* **100**, 467–472
19. Asahi, M., Sugita, Y., Kurzydlowski, K., De Leon, S., Tada, M., Toyoshima, C. & MacLennan, D.H. (2003) Sarcolipin regulates sarco(endo)plasmic reticulum Ca^{2+}-ATPase (SERCA) by binding to transmembrane helices alone or in association with phospholamban. *Proc. Natl. Acad. Sci. U.S.A.* **100**, 5040–5045
20. Kawase, Y. & Hajjar, R.J. (2008) The cardiac sarcoplasmic/endoplasmic reticulum calcium ATPase: a potent target for cardiovascular diseases. *Nat. Clin. Pract. Cardiovasc. Med.* **5**, 554–565
21. Vangheluwe, P., Sipido, K.R., Raeymaekers, L. & Wuytack, F. (2006) New perspectives on the role of SERCA2's Ca^{2+} affinity in cardiac function. *Biochim. Biophys. Acta* **1763**, 1216–1228
22. Haghighi, K., Schmidt, A.G., Hoit, B.D., Brittsan, A.G., Yatani, A., Lester, J.W., Zhai, J., Kimura, Y., Dorn, 2nd, G.W., MacLennan, D.H. & Kranias, E.G. (2001) Superinhibition of sarcoplasmic reticulum function by phospholamban induces cardiac contractile failure. *J. Biol. Chem.* **276**, 24145–24152
23. Schmitt, J.P., Kamisago, M., Asahi, M., Li, G.H., Ahmad, F., Mende, U., Kranias, E.G., MacLennan, D.H., Seidman, J.G. & Seidman, C.E. (2003) Dilated cardiomyopathy and heart failure caused by a mutation in phospholamban. *Science* **299**, 1410–1413
24. Haghighi, K., Kolokathis, F., Gramolini, A.O., Waggoner, J.R., Pater, L., Lynch, R.A., Fan, G.C., Tsiapras, D., Parekh, R.R., Dorn, 2nd, G.W. et al. (2006) A mutation in the human phospholamban gene, deleting arginine 14, results in lethal, hereditary cardiomyopathy. *Proc. Natl. Acad. Sci. U.S.A.* **103**, 1388–1393
25. Haghighi, K., Kolokathis, F., Pater, L., Lynch, R.A., Asahi, M., Gramolini, A.O., Fan, G.C., Tsiapras, D., Hahn, H.S., Adamopoulos, S. et al. (2003) Human phospholamban null results in lethal dilated cardiomyopathy revealing a critical difference between mouse and human. *J. Clin. Invest.* **111**, 869–876
26. Luo, W., Grupp, I.L., Harrer, J., Ponniah, S., Grupp, G., Duffy, J.J., Doetschman, T. & Kranias, E.G. (1994) Targeted ablation of the phospholamban gene is associated with markedly enhanced myocardial contractility and loss of β-agonist stimulation. *Circ. Res.* **75**, 401–409

27. Shanmugam, M., Gao, S., Hong, C., Fefelova, N., Nowycky, M.C., Xie, L.H., Periasamy, M. & Babu, G.J. (2011) Ablation of phospholamban and sarcolipin results in cardiac hypertrophy and decreased cardiac contractility. *Cardiovasc. Res.* **89**, 353–361
28. Jaski, B.E., Jessup, M.L., Mancini, D.M., Cappola, T.P., Pauly, D.F., Greenberg, B., Borow, K., Dittrich, H., Zsebo, K.M. & Hajjar, R.J. (2009) Calcium upregulation by percutaneous administration of gene therapy in cardiac disease (CUPID Trial), a first-in-human phase 1/2 clinical trial. *J. Card. Fail.* **15**, 171–181
29. Gwathmey, J.K., Yerevanian, A.I. & Hajjar, R.J. (2011) Cardiac gene therapy with SERCA2a: from bench to bedside. *J. Mol. Cell. Cardiol.*, doi:10.1016/j.yjmcc.2010.11.011
30. Suckau, L., Fechner, H., Chemaly, E., Krohn, S., Hadri, L., Kockskamper, J., Westermann, D., Bisping, E., Ly, H., Wang, X. et al. (2009) Long-term cardiac-targeted RNA interference for the treatment of heart failure restores cardiac function and reduces pathological hypertrophy. *Circulation* **119**, 1241–1252
31. Hoshijima, M., Ikeda, Y., Iwanaga, Y., Minamisawa, S., Date, M.O., Gu, Y., Iwatate, M., Li, M., Wang, L., Wilson, J.M. et al. (2002) Chronic suppression of heart-failure progression by a pseudophosphorylated mutant of phospholamban via *in vivo* cardiac rAAV gene delivery. *Nat. Med.* **8**, 864–871
32. Toyoshima, C., Nakasako, M., Nomura, H. & Ogawa, H. (2000) Crystal structure of the calcium pump of sarcoplasmic reticulum at 2.6 Å resolution. *Nature* **405**, 647–655
33. Morth, J.P., Pedersen, B.P., Toustrup-Jensen, M.S., Sorensen, T.L., Petersen, J., Andersen, J.P., Vilsen, B. & Nissen, P. (2007) Crystal structure of the sodium–potassium pump. *Nature* **450**, 1043–1049
34. Shinoda, T., Ogawa, H., Cornelius, F. & Toyoshima, C. (2009) Crystal structure of the sodium–potassium pump at 2.4 Å resolution. *Nature* **459**, 446–450
35. Pedersen, B.P., Buch-Pedersen, M.J., Morth, J.P., Palmgren, M.G. & Nissen, P. (2007) Crystal structure of the plasma membrane proton pump. *Nature* **450**, 1111–1114
36. Toyoshima, C. (2009) How Ca^{2+}-ATPase pumps ions across the sarcoplasmic reticulum membrane. *Biochim. Biophys. Acta* **1793**, 941–946
37. Moller, J.V., Olesen, C., Winther, A.L. & Nissen, P. (2010) The sarcoplasmic Ca^{2+}-ATPase: design of a perfect chemi-osmotic pump. *Q. Rev. Biophys.* **43**, 501–566
38. Vangheluwe, P., Sepulveda, M.R., Missiaen, L., Raeymaekers, L., Wuytack, F. & Vanoevelen, J. (2009) Intracellular Ca^{2+}- and Mn^{2+}-transport ATPases. *Chem. Rev.* **109**, 4733–4759
39. Vandecaetsbeek, I., Trekels, M., De Maeyer, M., Ceulemans, H., Lescrinier, E., Raeymaekers, L., Wuytack, F. & Vangheluwe, P. (2009) Structural basis for the high Ca^{2+} affinity of the ubiquitous SERCA2b Ca^{2+} pump. *Proc. Natl. Acad. Sci. U.S.A.* **106**, 18533–18538
40. Verboomen, H., Wuytack, F., Van Den Bosch, L., Mertens, L. & Casteels, R. (1994) The functional importance of the extreme C-terminal tail in the gene 2 organellar Ca^{2+}-transport ATPase (SERCA2a/b). *Biochem. J.* **303**, 979–984
41. Dode, L., Andersen, J.P., Leslie, N., Dhitavat, J., Vilsen, B. & Hovnanian, A. (2003) Dissection of the functional differences between sarco(endo)plasmic reticulum Ca^{2+}-ATPase (SERCA) 1 and 2 isoforms and characterization of Darier disease (SERCA2) mutants by steady-state and transient kinetic analyses. *J. Biol. Chem.* **278**, 47877–47889
42. John, L.M., Lechleiter, J.D. & Camacho, P. (1998) Differential modulation of SERCA2 isoforms by calreticulin. *J. Cell Biol.* **142**, 963–973
43. Roderick, H.L., Lechleiter, J.D. & Camacho, P. (2000) Cytosolic phosphorylation of calnexin controls intracellular Ca^{2+} oscillations via an interaction with SERCA2b. *J. Cell Biol.* **149**, 1235–1248
44. Hasler, U., Crambert, G., Horisberger, J.D. & Geering, K. (2001) Structural and functional features of the transmembrane domain of the Na,K-ATPase β subunit revealed by tryptophan scanning. *J. Biol. Chem.* **276**, 16356–16364
45. Vandecaetsbeek, I., Christensen, S.B., Liu, H., Van Veldhoven, P.P., Waelkens, E., Eggermont, J., Raeymaekers, L., Moller, J.V., Nissen, P., Wuytack, F. & Vangheluwe, P. (2011) Thapsigargin affinity purification of intracellular P_{2A}-type Ca^{2+}-ATPases. *Biochim. Biophys. Acta*, doi:10.1016/j.bbamcr.2010.12.020
46. Sagara, Y. & Inesi, G. (1991) Inhibition of the sarcoplasmic reticulum Ca^{2+} transport ATPase by thapsigargin at subnanomolar concentrations. *J. Biol. Chem.* **266**, 13503–13506
47. Winther, A.M., Liu, H., Sonntag, Y., Olesen, C., le Maire, M., Soehoel, H., Olsen, C.E., Christensen, S.B., Nissen, P. & Moller, J.V. (2010) Critical roles of hydrophobicity and orientation of side chains for inactivation of sarcoplasmic reticulum Ca^{2+}-ATPase with thapsigargin and thapsigargin analogs. *J. Biol. Chem.* **285**, 28883–28892
48. Takahashi, M., Kondou, Y. & Toyoshima, C. (2007) Interdomain communication in calcium pump as revealed in the crystal structures with transmembrane inhibitors. *Proc. Natl. Acad. Sci. U.S.A.* **104**, 5800–5805

Biochem. Soc. Symp. 78
Citation reference: Biochem. Soc. Trans. (2011) **39**, 789–797.

11

A diversity of SERCA Ca^{2+} pump inhibitors

Francesco Michelangeli*[1] and J. Malcolm East†

**School of Biosciences, University of Birmingham, Edgbaston, Birmingham B15 2TT, U.K., and †School of Biological Sciences, Life Sciences Building, University of Southampton, Highfield, Southampton SO17 1BJ, U.K.*

Abstract

The SERCA (sarcoplasmic/endoplasmic reticulum Ca^{2+}-ATPase) is probably the most extensively studied membrane protein transporter. There is a vast array of diverse inhibitors for the Ca^{2+} pump, and many have proved significant in helping to elucidate both the mechanism of transport and gaining conformational structures. Some SERCA inhibitors such as thapsigargin have been used extensively as pharmacological tools to probe the roles of Ca^{2+} stores in Ca^{2+} signalling processes. Furthermore, some inhibitors have been implicated in the cause of diseases associated with endocrine disruption by environmental pollutants, whereas others are being developed as potential anticancer agents. The present review therefore aims to highlight some of the wide range of chemically diverse inhibitors that are known, their mechanisms of action and their binding location on the Ca^{2+} ATPase. Additionally, some ideas for the future development of more useful isoform-specific inhibitors and anticancer drugs are presented.

[1]*To whom correspondence should be addressed (email F.Michelangeli@bham.ac.uk).*

Introduction

For more than 40 years, there has been considerable research into molecules that can affect the activity of one of the most extensively studied membrane protein transporters, the SERCA (sarcoplasmic/endoplasmic reticulum Ca^{2+}-ATPase). Originally identified in muscle in 1962 by Ebashi and Ebashi [1], this easy to isolate and study (at least in the skeletal muscle) membrane protein transporter has been the focus of considerable scientific attention over the intervening years. Initially, inhibitor studies were used to help to elucidate mechanistic and kinetic details of this Ca^{2+}-ATPase; however, more recently, the use of inhibitors has aided our determination of distinct conformational structures using X-ray crystallography of SERCA–inhibitor complexes.

Some inhibitors, such as thapsigargin, BHQ [2,5-di-(t-butyl)-1,4-hydroquinone] and CPA (cyclopiazonic acid), have also become widely employed as pharmacological tools in the cell signalling field. For instance, to date, more than 7000 scientific papers have been published that have used thapsigargin in their investigations of various Ca^{2+} signalling properties in cells and tissues. More recently, inhibitors of SERCA pumps in *Plasmodium falciparum*, are also being tested as pharmaceutical drugs to combat malaria [2]. Focus has also been given to identifying isoform-specific SERCA inhibitors, since it has been shown that some types of cancer cell appear to have altered expression of particular SERCA isoforms [3,4] and this could therefore lead to the development of novel anticancer drugs [5].

A diversity of Ca^{2+}-ATPase inhibitors

To date, many (probably hundreds) SERCA inhibitors with a variety of chemical structures have been identified. They range from small, but unrelated, hydrophobic molecules, which may contain hydroxy groups, to small charged anions such as orthovanadate or more complex peptide toxins such as mastoparan (Figure 1). Many of these diverse inhibitor molecules are able to inhibit the Ca^{2+}-ATPase in the low-micromolar to nanomolar concentration ranges, indicating high binding affinities. However, given such high-affinity binding, it is clear from the Ca^{2+}-ATPase structure that there are only a limited number of potential binding sites on the protein where these diverse molecules can bind.

Thapsigargin

As highlighted above, thapsigargin is by far the most widely used SERCA inhibitor. It was initially identified from an extract from the plant *Thapsia garganica* that could increase free cytosolic Ca^{2+} levels in platelets [6], and this was later determined to be due to inhibition of the endoplasmic reticulum Ca^{2+}-ATPase [7]. Thapsigargin is a sesquiterpene lactone that can inhibit SERCA in the nanomolar concentration range [7,8]. It is also highly selective, since it does not appreciably inhibit other related Ca^{2+}-ATPases such as the PMCA (plasma

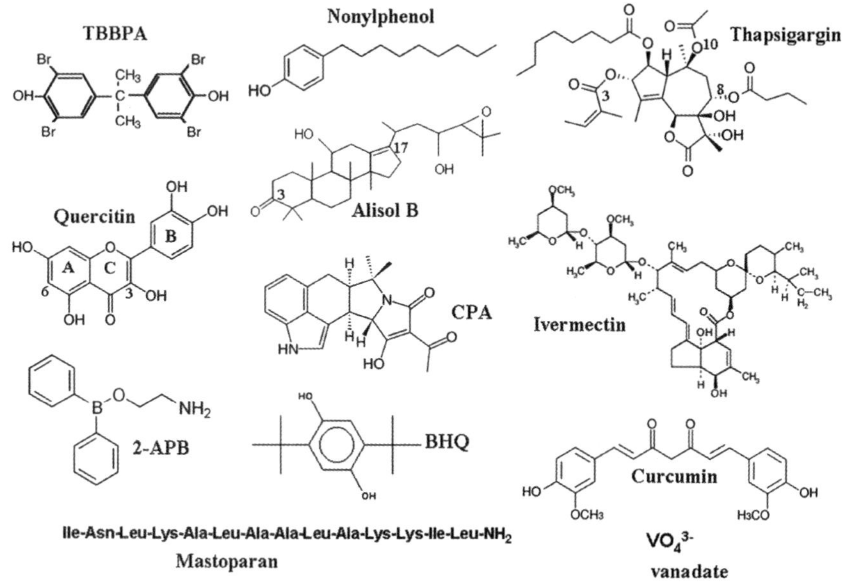

Figure 1. A diverse range of SERCA inhibitors

membrane Ca^{2+}-ATPase) or the SPCA (secretory pathway Ca^{2+}-ATPase), at these very low concentrations [9,10]. Furthermore, it appears that thapsigargin also differentially inhibits SERCA isoforms, being 60 times more potent for SERCA1 than for SERCA3 (K_i values of 0.2, 1 and 12 nM for SERCA-isoforms 1, 2 and 3 respectively) [8]. Initial enzymology studies suggested that thapsigargin caused inhibition by both inhibiting Ca^{2+} binding and phosphorylation [11]. Later studies additionally showed that it did so by stabilizing the E2 (low Ca^{2+} affinity) conformational state causing it to be locked into an E2-type 'dead-end' state that was practically irreversible [12]. The fact that thapsigargin could lock it in a E2-type conformation was also instrumental in enabling the E2 conformational state crystal structure of SERCA to be determined [13].

The use of thapsigargin-insensitive mutant cells initially identified Phe^{256} within SERCA1 as an amino acid residue residing in the thapsigargin-binding site, since mutation of this residue to valine (F256V) resulted in a 200-fold decrease in thapsigargin sensitivity [8]. Interestingly, all other SERCA isoforms contained Phe^{256}, which, when mutated, reduced their sensitivities to thapsigargin, albeit to lesser degrees than SERCA1 [8].

In 2002, the crystal structure of SERCA1 in the E2 state, in which thapsigargin was bound, identified the exact binding pocket that involved interactions with transmembrane helices M3, M5 and M7. Glu^{255}, Phe^{256}, Gln^{259}, Leu^{260}, Val^{263}, Val^{769}, Ile^{765}, Phe^{834}, Met^{838} and Tyr^{837} are particularly important in forming hydrogen bonds and other interactions with thapsigargin [13,14] (Figure 2). More recently, much attention has been given to the making

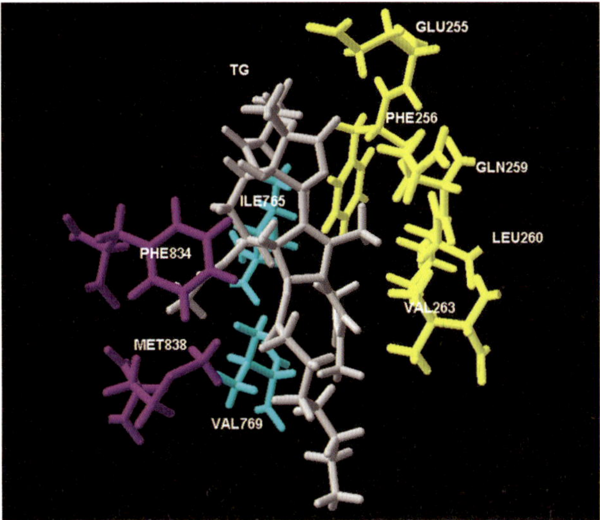

Figure 2. The thapsigargin-binding site
Close-up of the thapsigargin-binding site within SERCA1A in the E2 conformational state. Highlighted are a number of the key amino acid residues which are believed to be important for binding. Amino acids shown in yellow are located within transmembrane helix M3, those in cyan are in M5 and those in purple are in M7. The crystal structure was obtained from PBD code 2AGV.

and testing of structural analogues of thapsigargin in order to better understand the molecular interactions between thapsigargin and SERCA. These studies have shown that the acyl groups at positions O-3, O-8 and O-10 are particularly important, probably due to their appropriate hydrophobic interactions with SERCA [14]. It must be noted that most, if not all, thapsigargin analogues so far tested are weaker inhibitors than the parent compound. It has also been proposed that thapsigargin is only able to gain access to its buried hydrophobic binding site within the ATPase by first partitioning into the lipid membrane rather than gaining direct access from the aqueous phase [14].

In screening a number of natural products extracted from traditional Chinese medicinal herbs, in order to discover novel autophagy enhancers that could potentially be used as therapeutic agents, a compound called Alisol B was identified [15]. In addition to causing autophagy and cell death in a number of cancer cell lines, this compound was also shown to elevate intracellular Ca^{2+} levels via SERCA inhibition (IC_{50} 27 μM) [15]. The structure of Alisol B is steroid in nature with a ketone group at the C-3 position and a branched acyl hydroxylated epoxide side chain at the C-17 position (Figure 1) [15]. Energy-minimized molecular docking analysis showed that Alisol B was likely to bind best to the transmembrane domain at the same site occupied by thapsigargin. The best binding pose was achieved when the steroid ring was in close contact with Phe^{256}, Val^{263}, Ile^{765} and Val^{769}, all very similar to the interactions seen with thapsigargin [14,15].

Table 1. Potencies and conformational effects of some SERCA inhibitors

Inhibitor	K_i or IC_{50}	E1-like or E2-like conformation	Reference(s)
Thapsigargin	0.21–12 nM	E2	[8]
Cyclopiazonic acid	90–2500 nM	E2	[8]
BHQ	2–7 μM	E2	[8]
Nonylphenol	6 μM	E2	[22,31]
Bisphenol	2 μM	E2	[33]
Bisphenol A	233 μM	E2?	[23]
TBBPA	0.5–2.3 μM	E2	[30]
4-Chloro-*m*-cresol	2.8 mM	?	[54]
Orthovanadate	10–100 μM	E2	[55]
Quercitin	9 μM	E1	[40]
3,6-Dihydroxyflavone	6 μM	E1	[40]
Galangin	9 μM	E1	[40]
2APB	70–700 μM	E1	[43]
Curcumin	7–15 μM	E1	[35]
Paxilline	5 μM	E1	[56]
Alisol B	27 μM	E2	[15]
Mastoparan	1 μM	E1	[57,58]
Peptide M391	0.3 μM	E1?	[57,58]
Ivermectin	15 μM	E1	[59]
Cyclosporin A	62 μM	?	[59]
Rapamycin	77 μM	?	[59]
Chlorpromazine	23 μM	E1	[60]
Calmidazolium	0.5 μM	E1	[60]
Fluphenazine	15 μM	E1	[60]
DES	20 μM	?	[23]
1,3-Dibromo-2,4,6-tris(methyliso-thiouronium)benzene (Br_2-TITU)	29 μM	E1	[32]
sHA 14-1	23 μM	?	[61]

Early studies with cholesterol and androstenol showed that they do not appear to affect the activity of SERCA1 when added to it in its native-like phospholipid membrane [15,17]. They do, however, bind directly to SERCA, as demonstrated by fluorescence quenching studies, where the intrinsic typtophan fluorescence of the Ca^{2+} ATPase is quenched by contact with brominated sterol analogues [16,17]. These sites were also shown not be at the lipid–protein interface where annular lipids bind, since additional quenching was observed when the ATPase was first reconstituted in brominated phospholipids. From these studies, the existence of hydrophobic 'non-annular' binding sites on the Ca^{2+}-ATPase was first proposed [16]. When such sites are occupied by these sterols in SERCA which has been reconstituted in suboptimal fatty acyl chain length phospholipids, a dramatic enhancement in activity is observed [16, 17]. Therefore these sterol-binding sites, which have also been observed in other ion-translocating ATPases [18], may be the evolutionary remnants of some lost regulatory mechanism. Detailed structural analysis of the thapsigargin-binding site has led to the suggestion that this site and the 'non-annular' sites could be one and the same [14]; however, this has yet to be proven.

Inhibitors that stabilize the E2 conformation

Table 1 lists some of the vast array of diverse compounds that are able to act as SERCA inhibitors and included in this Table are inhibition constants (or IC_{50}) values for these inhibitors. Some of these molecules, such as bisphenol A, nonylphenol and DES (diethylstilbestrol), have been in the limelight recently owing to their endocrine-disrupting properties which are linked with male infertility, and which are believed to be due to their ability to bind to oestrogen receptors [19,20]. However, it cannot be discounted that some of these chemicals which pervade our environment could also be acting as endocrine disrupters by dysregulating Ca^{2+} signalling events through their action on SERCA [21–23]. Furthermore, of note are BHQ and CPA, which, like thapsigargin, have been used extensively as pharmacological tools by scientists to mobilize SERCA-loaded Ca^{2+} stores [24]. Both inhibitors were shown previously to inhibit the Ca^{2+}-ATPase by stabilizing it in one of the E2 conformational states [12,25] and, to date, many other inhibitors of this class have also been shown to cause this same effect. It was originally proposed that molecules such as CPA and BHQ bound to the same site as thapsigargin [26]; however, this was challenged by inhibitor competition studies [27], and then confirmed by crystal structures clearly showing them binding to different sites [28,29]. One crystal structure was produced containing both bound thapsigargin and BHQ (PDB code 2AGV) which clearly shows these two sites on opposite sides of the transmembrane bundle (Figure 3). From these structures, it appears that in the E2-like conformational state a buried hydrophobic groove is created between transmembrane helices M1, M2, M3 and M4, owing to M1 adopting a 'kinked' conformation driven by movement of the actuator domain. This allows access of BHQ or CPA to bind to this site [28,29]. Amino acids important in inhibitor binding to this site include a number of polar ones such as Gln^{56}, Asp^{59} and Asn^{101}, which presumably interact with the hydroxy groups on these inhibitors [28,29]. Fluorescence-quenching studies using the brominated hydrophobic inhibitor TBBPA (tetrabromobisphenol A) causes quenching of the SERCA tryptophan fluorescence and can be displaced by BHQ, but not by thapsigargin. This therefore indicates that TBBPA also binds to the BHQ-binding site [30]. Detailed mechanistic studies have shown that TBBPA also stabilizes the enzyme in an E2 state [30]. Additionally, nonylphenol which also stabilizes the E2 state [31], can displace TBBPA and reverse fluorescence quenching [30], and it therefore appears likely that this and a number of other small hydroxylated hydrophobic inhibitors (which lock the enzyme in a E2 state) will work in a similar fashion.

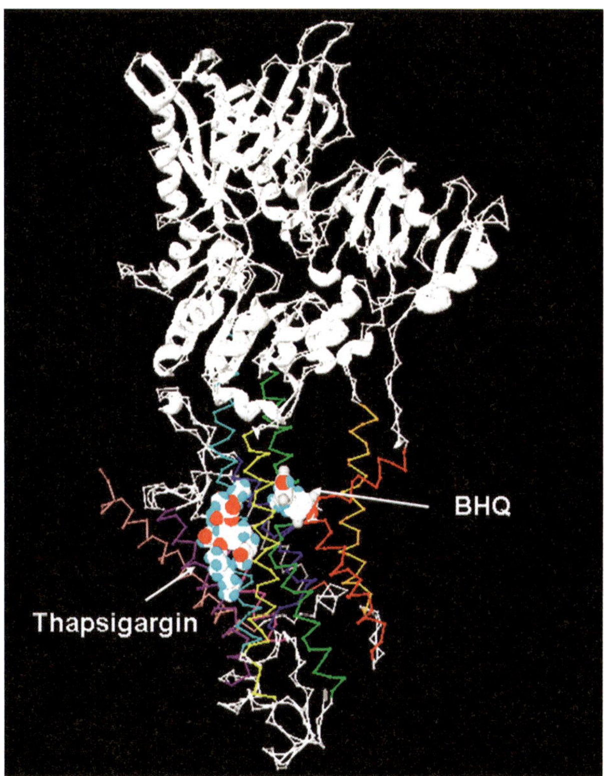

Figure 3. Location of thapsigargin and BHQ inhibitor-binding sites in SERCA1A
Shown is the SERCA1A crystal structure in which thapsigargin and BHQ are bound (PBD code 2AGV). The two binding sites are distinct and are at opposites sides of the transmembrane bundle. The transmembrane helices are colour-coded accordingly: M1 (red), M2 (orange), M3 (yellow), M4 (green), M5 (cyan), M6 (blue), M7 (purple), M8 (pink), M9 (grey) and M10 (brown).

One can envisage these small molecules acting as a 'doorstop', wedging open the gap between the helix M1 and helices M2, M3 and M4, keeping it ajar and thus making it unable to go back into the E1 conformation. The region where TBBPA binds is also postulated to be very close to Ca^{2+}-binding site II, which could also explain how TBBPA reduces the Ca^{2+}-binding affinity more than 20-fold [30].

The use of brominated inhibitors such as TBBPA have also been useful in determining whether this class of inhibitor can either partition directly from the aqueous phase into the Ca^{2+}-ATPase or indirectly by first partitioning into the lipid phase. By measuring the rate constants for TBBPA binding to the Ca^{2+}-ATPase (by following its ability to quench protein fluorescence), it was found that the rates were approximately 10-fold higher than for TBBPA binding to phospholipid bilayers alone, indicating that this inhibitor is able to bind directly to SERCA [30].

Inhibitors that stabilize the E1 conformational state

Although it is clear from both enzymatic and structural studies that many of these small-molecule hydrophobic inhibitors bind and stabilize the Ca^{2+}-ATPase in the E2-like state, by binding to one of the two sites highlighted above, more recent studies have also highlighted hydrophobic inhibitors which can bind to SERCA in the E1-like conformation. These E1 inhibitors are therefore likely to bind at other sites on the protein. Table 1 lists some of these hydrophobic inhibitors and as can be seen their potency for inhibition varies from the micromolar to millimolar range. A number of approaches have been used to assess E1 stabilization, including monitoring the E1P-Ca^{2+}/E2P-Ca^{2+} ratio by assessing ADP-sensitive to ADP-insensitive phosphoenzyme levels [32], monitoring the E1–E2 step by employing fluorescently labelled ATPase [12,33], or by analysing conformational-dependent proteolytic digestion [32,34].

One of the first inhibitors to be conclusively shown to inhibit the Ca^{2+}-ATPase by stabilizing the E1 conformation was curcumin [1,7-bis(4-hydroxy-3-methoxyphenol)-1,6-heptadiene-3,5-dione] [35]. Curcumin is derived from the spice tumeric, which has been used as an anti-inflammatory agent and is currently being investigated as an anticancer drug [36]. Curcumin inhibits SERCA1 activity with a K_i of 15 μM. This inhibition is non-competitive with respect to Ca^{2+} and competitive with respect to ATP, as determined by a reduction in both ATP binding and ATP-dependent phosphoenzyme formation [35]. In addition, experiments with FITC-labelled ATPase show that it stabilizes the E1 conformation. FITC is known to label the Ca^{2+}-ATPase at Lys^{515} within the ATP-binding site. Labelling this position of the ATPase with FITC is known to inhibit ATP binding. The fact that curcumin still alters FITC–ATPase fluorescence must indicate that it does not bind to the nucleotide-binding site directly, but rather to another site within the ATPase that then induces a conformational change to prevent ATP binding. These findings have been interpreted as curcumin stabilizing the interaction between the nucleotide-binding and phosphorylation domains inhibiting ATP from binding. In order for phosphorylation to occur the nucleotide-binding domain (with ATP bound) and the phosphorylation domain need to come into close contact. Both domains are highly mobile and are known to move together and undergo major rearrangements in going from E1 to E2 conformations [13,37]. Additionally, in the E1 conformations, these two domains can also come into contact with each other by simple thermal fluctuations [37]. As these two domains are linked together via a 'hinge' region, we have speculated that curcumin could affect this region of the ATPase, locking the two domains together and thereby occluding the ATP-binding site. Alternatively, curcumin may stabilize the association between the nucleotide and phosphorylation domains by binding to the interface.

Flavonoids are commonly found in fruit and vegetables and have been shown to reach concentrations of several micromolar on human blood plasma [38]. Flavonoids are heterocyclic compounds consisting of three linked rings,

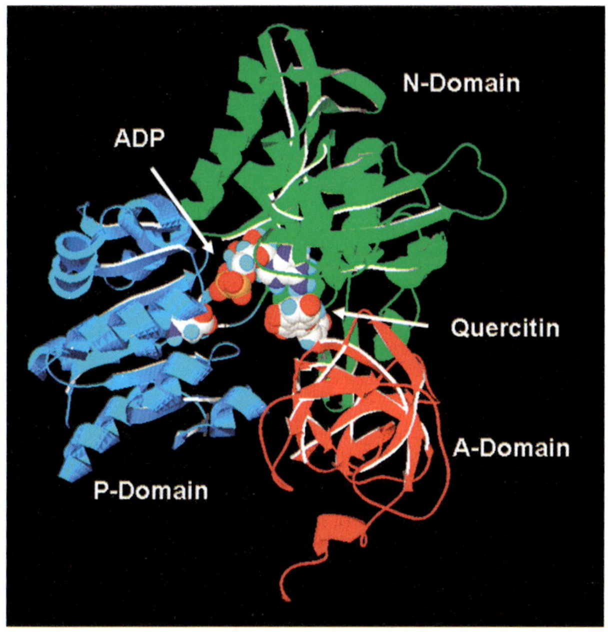

Figure 4. Putative location of the flavonoid-binding site
The Molegro Virtual Docker 2007 program was used to predict the flavonoid (quercitin)-binding site on the E1. Ca^{2+}-ADP-bound structure of SERCA1A (PDB code 1WPE). The binding site was predicted to occur at the interface of the nucleotide (N-, green), phosphorylation (P-, blue) and actuator (A-, red) domains. The ADP-binding site (which is shown occupied) could become occluded when quercitin has bound, and this would likely affect ATP binding and phosphorylation.

of which two are aromatic (see Figure 1). These compounds are also believed to have cancer chemoprotective properties, possibly by triggering apoptosis via the Ca^{2+}-dependent mitochondrial pathway [39]. One mechanism by which this can occur is via exaggerated increases in cytosolic [Ca^{2+}], which could be due to the fact that some flavonoids are able to potently inhibit SERCA [40]. Of an extensive range of flavonoids tested, the most potent inhibitors were 3,6-dihydroxyflavone, quercetin and galangin (Table 1) [40]. A quantitative structure–activity relationship study indicated that polyhydroxylation was important, with hydroxylation at positions 3 and 6 (on rings C and A respectively). A detailed study also showed that the mechanism by which 3,6-dihydroxyflavone and galangin inhibit SERCA1A appears to be by altering the ATP affinity and the associated ATP-dependent phosphorylation step, in addition to stabilizing the enzyme in an E1 conformational state. Again, using energy-minimization molecular modelling programs, it was shown that these flavonoids appear to bind to the cytosolic region of the Ca^{2+}-ATPase between the ATP-binding and phosphorylation domains (Figure 4). Thus these flavonoids could prevent ATP from binding, in a similar manner to that postulated for curcumin, rather than directly occupying the nucleotide-binding site.

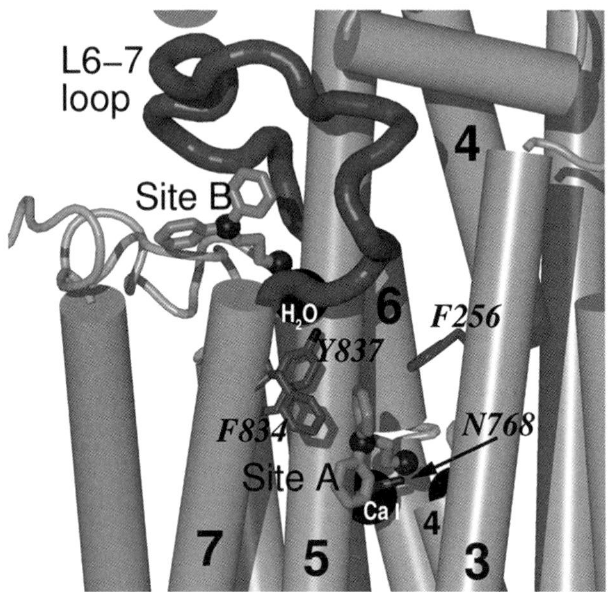

Figure 5. Putative 2APB-binding sites
The two possible sites of interaction for 2-APB with the E1 form of the ATPase (PBD code 1EUL) that was predicted in [43] are shown. The two sites are either with the L6–L7 loop or between transmembrane helices M3 and M5.

2APB (2-aminoethoxydiphenyl borate) has been used as both an InsP_3 receptor inhibitor [41] and an activator of store-operated Ca^{2+} entry [42]. However, our research has also shown it to be a pH-sensitive SERCA inhibitor with an IC_{50} of between 70 μM at pH 6 and 700 μM at pH 7 [43]. Although a relatively weak inhibitor, it was shown to inhibit Ca^{2+} binding 20-fold, reducing the association rate of binding and increasing its rate of dissociation [43]. Furthermore, 2APB reduced phosphoryl transfer without inhibiting ATP binding, but still stabilizing the E1 conformational state [43]. Activity studies using a mutant form of SERCA1 where Tyr^{837} was replaced by phenylalanine showed that it became insensitive to inhibition with 2APB. Mutation of another amino acid which is in close proximity to this residue (F834A), however, had little effect on 2-APB's ability to inhibit SERCA1 [43]. Molecular modelling studies identified two potential binding sites for 2APB close to this residue near to transmembrane helices M3, M4, M5 and M7 and close to the cytoplasmic loop between M6 and M7 (L6–L7) (Figure 5). Taking into account that our studies showed that 2APB inhibited Ca^{2+} binding, with the proposal that the L6–L7 loop has been implicated in the Ca^{2+} entry pathway/route by which Ca^{2+} can enter their Ca^{2+}-binding sites [44,45], led us to suggest that, when 2APB binds to either of these sites, it acts as a 'plug' blocking access to Ca^{2+}-binding site I [43].

Can inhibitor studies tell us about the Ca^{2+}-entry pathways?

There have so far been several proposed Ca^{2+}-entry pathways by which the two Ca^{2+} ions gain access to their respective binding sites [37,43,46,]; however, there is no clear consensus. Our studies have so far identified two inhibitors (2APB and TBBPA) which can both inhibit Ca^{2+} binding via different mechanisms involving different conformational states (E1 for 2APB and E2 for TBBPA). Furthermore, activity, mutagenesis and modelling studies have predicted that each bind to different sites at opposite sides of the transmembrane bundle, close to Ca^{2+}-binding site I in the case of 2APB and Ca^{2+}-binding site II for TBBPA. These observations have led us to speculate that the two Ca^{2+} ions bind to their respective binding sites by different routes located at adjacent sides of the transmembrane bundle [30]. If such a possibility exists, then one would predict that, under some circumstances, the stoichiometry for Ca^{2+} binding can go from 2 to 1 if one pathway was blocked independently. In support of this, we have already shown that when the Ca^{2+}-ATPase is reconstituted into short-chain phospholipid bilayers, the stoichiometry for Ca^{2+} binding is reduced from 2 to 1 [47]. However, to confirm this hypothesis, more detailed Ca^{2+}-binding, mutagenesis and structural studies need to be undertaken.

Future directions

Isoform-specific inhibitors

Three isoforms of SERCA are known to exist in mammals, with approximately 75 % sequence homology between one another and each isoform also existing in a variety of splice variant forms [48]. The expression profiles of the SERCA isoforms are known to vary both in different tissues and during development and disease [3,4,48]. Most tissue/cell types appear to express more than one SERCA isoform or splice-variant type, with each potentially associated with physiologically distinct Ca^{2+} stores [49]. It would therefore be highly desirable to be able to distinguish and study these SERCA-isoform-specific Ca^{2+} stores using isoform-specific inhibitors. A comparative study of the effects of a number of SERCA inhibitors was undertaken using cells overexpressing each isoform subtype. As highlighted above, thapsigargin is the most potent of all inhibitors, and its potency is different for the three isoforms, being most potent for SERCA1 ($K_i = 0.2$ nM) and relatively least potent for SERCA3 ($K_i = 12$ nM). Modelling of the three-dimensional structures of SERCA2 and SERCA3 suggested that there were minor differences in the thapsigargin-binding sites within the three isoforms which could account for these differences in potency [8]. CPA also showed some isoform selectivity, again being most potent for SERCA1 ($K_i =$ 90 nM), least potent for SERCA2b ($K_i = 2.5$ μM) and intermediate for SERCA3 ($K_i = 600$ nM). All of the other inhibitors tested in this study showed only minor differences [8], except for curcumin, which was 6-fold less effective in inhibiting SERCA3a compared with SERCA2b. As most non-muscle tissues express either

SERCA2b alone or in combination with SERCA 3, it would be ideal if one could identify better inhibitors which could be highly selective between these two isoforms. Therefore investigating the selective potencies of curcumin analogues may well prove useful.

Inhibitors that can differentiate between SERCA, PMCA and SPCA

It is clear that some SERCA inhibitors such as vanadate can also inhibit PMCAs and SPCA located on the Golgi membrane. In order to selectively distinguish between these different types of Ca^{2+}-ATPases, thapsigargin has been the pharmacological tool of choice. It can be used in the 10–1000 nM range on cells to selectively inhibit SERCA without unduly affecting PMCA and SPCA, which requires several micromolar to achieve any degree of inhibition.

An approach involving screening a random peptide library for the binding of peptides to PMCA extracellular domains has recently led to the discovery of novel peptide PMCA inhibitors called caloxins which are highly selective towards PMCA over SERCA [50]. One such inhibitor, caloxin 1c2 (TAWSEVLDLLRRGGGSK-amide), has a K_i of 2.3 μM for PMCA4. Thus these or related peptides may well prove to be useful in differentiating the roles that PMCA and SERCA play in cellular Ca^{2+} homoeostasis.

As yet, there are no inhibitors which can selectively inhibit SPCA. However, such an inhibitor would prove very useful in investigating the role SPCA and SPCA-loaded Ca^{2+} stores play in protein processing and protein trafficking through the Golgi.

Inhibitors as cancer therapeutic agents

Owing to the major role SERCA plays in maintaining intracellular Ca^{2+} levels within acceptable limits in order to avoid Ca^{2+}-mediated autophagy and apoptosis, there is some interest in its role in diseases such as cancer [3–5]. It has been observed in some transformed cells such as colon epithelial cells that there is a down-regulation of expression of SERCA3 when the cells become transformed during carcinogenesis [3, 51]. It also appears that the decreased expression of SERCA3 is more pronounced the more undifferentiated the cells become [3]. Conversely, colon cancer cells such as CaCO-2 can be induced to differentiate and show signs of more normal phenotypic behaviour if allowed to become highly confluent in culture. Under such conditions, they also increase their SERCA3 expression levels, proportionately [3]. Decreased expression of SERCA3 was also noted in transformed gastric cells and lymphocytic cells; however, little or no changes in SERCA2 expression levels were noted in these three cell types in transformed and differentiated states [3]. Furthermore, mutations in SERCA3 have also recently been found to occur in some head and neck squamous cell carcinomas [52]. Cancer cell SERCA isoform expression profiles also appear to be cell-type-specific, since when comparing SERCA2b expression in normal and transformed thyroid cells, a dramatic down-regulation in the expression of this particular isoform was shown to occur [4].

The fact that SERCA is decreased in some transformed cells and that some cancers cells appear to be more prone to Ca^{2+}-mediated apoptosis has led to an approach of developing thapsigargin-based anticancer drugs. Thapsigargin analogues have been developed into prodrugs by coupling them to a targeting peptide to produce an inactive precursor that only becomes activated once the specific peptide sequence is cleaved by a tissue-specific protease. Currently, a thapsigargin prodrug with a peptide that is targeted by prostate-specific proteases is being evaluated as a potential cancer chemotherapeutic agent for the treatment of prostate cancer [5,53].

As highlighted above, a number of flavonoids and curcumin are also being evaluated as anticancer agents owing to their anti-proliferative or apoptotic properties [36,40]. It would now appear that these compounds could also be exerting their effects on cancer cells by triggering apoptosis through a Ca^{2+}-mediated pathway involving SERCA inhibition [40], in addition to any effects on other pathways.

Concluding remarks

Inhibitor studies on SERCA pumps were of paramount importance in helping to elucidate the mechanism by which this transporter works. More recently, inhibitors have been useful in helping to determine the tertiary structure of specific conformational states.

Over the years, it has become apparent that SERCAs play an important role in Ca^{2+} homoeostasis in all mammalian cells and that their inhibition can lead to cell death. In identifying potent SERCA inhibitors, we now have a novel strategy for targeting certain types of cancer cells. It is hoped that, in the future, more potent and specifically targeted inhibitors can be developed to combat this disease.

Acknowledgements

We thank Dr Jon Ride (School of Biosciences, University of Birmingham) for his help with binding predictions.

Funding

The Wellcome Trust is thanked for financial support.

References

1. Ebashi, F. & Ebashi, S. (1962) Removal of calcium and relaxation in actomyosin systems. *Nature* **194**, 378–379
2. Cardi, D., Pozza, A., Arnou, B., Marchal, E., Clausen, J.D., Andersen, J.P., Krishna, S., Møller, J.V., le Maire, M. & Jaxel, C. (2010) Purified E255L mutant SERCA1a and purified PfATP6 are sensitive to SERCA-type inhibitors but insensitive to artemisinins. *J. Biol. Chem.* **285**, 26406–26416

3. Gélébart, P., Kovács, T., Brouland, J.P., van Gorp, R., Grossmann, J., Rivard, N., Panis, Y., Martin, V., Bredoux, R., Enouf, J. & Papp, B. (2002) Expression of endomembrane calcium pumps in colon and gastric cancer cells: induction of SERCA3 expression during differentiation. *J. Biol. Chem.* **277**, 26310–26320
4. Pacifico, F., Ulianich, L., De Micheli, S., Treglia, S., Leonardi, A., Vito, P., Formisano, S., Consiglio, E. & Di Jeso, B. (2003) The expression of the sarco/endoplasmic reticulum Ca^{2+}-ATPases in thyroid and its down-regulation following neoplastic transformation. *J. Mol. Endocrinol.* **30**, 399–409
5. Denmeade, S.R. & Isaacs, J. T. (2005) The SERCA pump as a therapeutic target: making a "smart bomb" for prostate cancer. *Cancer Biol Ther.* **4**, 14–22
6. Ali, H., Christensen, S.B., Foreman, J.C., Pearce, F.L., Piotrowski, W. & Thastrup, O. (1985) The ability of thapsigargin and thapsigargicin to activate cells involved in the inflammatory response. *Br. J. Pharmacol.* **85**, 705–712
7. Thastrup, O., Cullen, P.J., Drøbak, B.K., Hanley, M.R. & Dawson, A.P. (1990) Thapsigargin, a tumor promoter, discharges intracellular Ca^{2+} stores by specific inhibition of the endoplasmic reticulum Ca^{2+}-ATPase. *Proc. Natl. Acad. Sci. U.S.A.* **87**, 2466–2470
8. Wootton, L.L. & Michelangeli, F. (2006) The effects of the phenylalanine 256 to valine mutation on the sensitivity of sarcoplasmic/endoplasmic reticulum Ca^{2+} ATPase (SERCA) Ca^{2+} pump isoforms 1, 2, and 3 to thapsigargin and other inhibitors. *J. Biol. Chem.* **281**, 6970–6976
9. Bokkala, S., el-Daher, S.S., Kakkar, V.V., Wuytack, F. & Authi, K.S. (1995) Localization and identification of Ca^{2+}ATPases in highly purified human platelet plasma and intracellular membranes: evidence that the monoclonal antibody PL/IM 430 recognizes the SERCA 3 Ca^{2+}ATPase in human platelets. *Biochem. J.* **306**, 837–842
10. Dode, L., Andersen, J.P., Vanoevelen, J., Raeymaekers, L., Missiaen, L., Vilsen, B. & Wuytack, F. (2006) Dissection of the functional differences between human secretory pathway Ca^{2+}/Mn^{2+}-ATPase (SPCA) 1 and 2 isoenzymes by steady-state and transient kinetic analyses. *J. Biol. Chem.* **281**, 3182–3189
11. Sagara, Y. & Inesi, G. (1991) Inhibition of the sarcoplasmic reticulum Ca^{2+} transport ATPase by thapsigargin at subnanomolar concentrations. *J. Biol. Chem.* **266**, 13503–13516
12. Wictome, M., Michelangeli, F., Lee, A.G. & East, J.M. (1992) The inhibitors thapsigargin and 2,5-di(tert-butyl)-1,4-benzohydroquinone favour the E2 form of the Ca^{2+},Mg^{2+}-ATPase. *FEBS Lett.* **304**, 109–113
13. Toyoshima, C. & Nomura, H. (2002) Structural changes in the calcium pump accompanying the dissociation of calcium. *Nature* **418**, 605–611
14. Winther, A.M., Liu, H., Sonntag, Y., Olesen, C., le Maire, M., Soehoel, H., Olsen, C.E., Christensen, S.B., Nissen, P. & Møller, J.V. (2010) Critical roles of hydrophobicity and orientation of side chains for inactivation of sarcoplasmic reticulum Ca^{2+}-ATPase with thapsigargin and thapsigargin analogs. *J. Biol. Chem.* **285**, 28883–28892
15. Law, B.Y., Wang, M., Ma, D.L., Al-Mousa, F., Michelangeli, F., Cheng, S.H., Ng, M.H., To, K.F., Mok, A.Y., Ko, R.Y. et al. (2010) Alisol B, a novel inhibitor of the sarcoplasmic/endoplasmic reticulum Ca^{2+} ATPase pump, induces autophagy, endoplasmic reticulum stress, and apoptosis. *Mol. Cancer Ther.* **9**, 718–730
16. Simmonds, A.C., East, J.M., Jones, O.T., Rooney, E.K., McWhirter, J. & Lee, A.G. (1982) Annular and non-annular binding sites on the ($Ca^{2+}+Mg^{2+}$)-ATPase. *Biochim. Biophys. Acta* **693**, 398–406
17. Michelangeli, F., East, J.M. & Lee, A.G. (1990) Structural effects on the interaction of sterols with the ($Ca^{2+}+Mg^{2+}$)-ATPase. *Biochim. Biophys. Acta* **1025**, 99–108
18. Shinoda, T., Ogawa, H., Cornelius, F. & Toyoshima, C. (2009) Crystal structure of the sodium-potassium pump at 2.4 Å resolution. *Nature* **459**, 446–450
19. Sumpter, J.P. (1998) Xenoendorine disrupters: environmental impacts. *Toxicol. Lett.* ***102–103***, 337–342
20. Guillette, Jr, L.J., Brock, J.W., Rooney, A.A. & Woodward, A.R. (1999) Serum concentrations of various environmental contaminants and their relationship to sex steroid concentrations and phallus size in juvenile American alligators. *Arch. Environ. Contam. Toxicol.* **36**, 447–455
21. Hughes, P.J., McLellan, H., Lowes, D.A., Kahn, S.Z., Bilmen, J.G., Tovey, S.C., Godfrey, R.E., Michell, R.H., Kirk, C.J. & Michelangeli, F. (2000) Estrogenic alkylphenols induce cell death by inhibiting testis endoplasmic reticulum Ca^{2+} pumps. *Biochem. Biophys. Res. Commun.* **277**, 568–574
22. Michelangeli, F., Ogunbayo, O.A., Wootton, L.L., Lai, P.F., Al-Mousa, F., Harris, R.M., Waring, R.H. & Kirk, C.J. (2008) Endocrine disrupting alkylphenols: structural requirements for their adverse effects on Ca^{2+} pumps, Ca^{2+} homeostasis and Sertoli TM4 cell viability. *Chem. Biol. Interact* **176**, 220–226
23. Kirk, C.J., Bottomley, L., Minican, N., Carpenter, H., Shaw, S., Kohli, N., Winter, M., Taylor, E.W., Waring, R.H., Michelangeli, F. & Harris, R.M. (2003) Environmental endocrine disrupters dysregulate estrogen metabolism and Ca^{2+} homeostasis in fish and mammals via receptor-independent mechanisms. *Comp. Biochem. Physiol. Part A Mol. Integr. Physiol.* **135**, 1–8

24. Darby, P.J., Kwan, C.Y. & Daniel, E.E. (1993) Use of calcium pump inhibitors in the study of calcium regulation in smooth muscle. *Biol. Signals* **2**, 293–304
25. Tadini-Buoninsegni, F., Bartolommei, G., Moncelli, M.R., Tal, D.M., Lewis, D. & Inesi, G. (2008) Effects of high-affinity inhibitors on partial reactions, charge movements, and conformational states of the Ca^{2+} transport ATPase (sarco-endoplasmic reticulum Ca^{2+} ATPase). *Mol. Pharmacol.* **73**, 1134–1140
26. Ma, H., Zhong, L., Inesi, G., Fortea, I., Soler, F. & Fernandez-Belda, F. (1999) Overlapping effects of S3 stalk segment mutations on the affinity of Ca^{2+}-ATPase (SERCA) for thapsigargin and cyclopiazonic acid. *Biochemistry* **38**, 15522–15527
27. Logan-Smith, M.J., East, J.M. & Lee, A.G. (2002) Evidence for a global inhibitor-induced conformation change on the Ca^{2+}-ATPase of sarcoplasmic reticulum from paired inhibitor studies. *Biochemistry* **41**, 2869–2875
28. Obara, K., Miyashita, N., Xu, C., Toyoshima, I., Sugita, Y., Inesi, G. & Toyoshima, C. (2005) Structural role of countertransport revealed in Ca^{2+} pump crystal structure in the absence of Ca^{2+}. *Proc. Natl. Acad. Sci. U.S.A.* **102**, 14489–14496
29. Laursen, M., Bublitz, M., Moncoq, K., Olesen, C., Møller, J.V., Young, H.S., Nissen, P. & Morth, J.P. (2009) Cyclopiazonic acid is complexed to a divalent metal ion when bound to the sarcoplasmic reticulum Ca^{2+}-ATPase. *J. Biol. Chem.* **284**, 13513–13518
30. Ogunbayo, O.A. & Michelangeli, F. (2007) The widely utilized brominated flame retardant tetrabromobisphenol A (TBBPA) is a potent inhibitor of the SERCA Ca^{2+} pump. *Biochem. J.* **408**, 407–415
31. Michelangeli, F., Orlowski, S., Champeil, P., East, J.M. & Lee, A.G. (1990) Mechanism of inhibition of the (Ca^{2+}-Mg^{2+})-ATPase by nonylphenol. *Biochemistry* **29**, 3091–3101
32. Hua, S., Xu, C., Ma, H. & Inesi, G. (2005) Interference with phosphoenzyme isomerization and inhibition of the sarco-endoplasmic reticulum Ca^{2+} ATPase by 1,3-dibromo-2,4,6-tris(methylisothiouronium) benzene. *J. Biol. Chem.* **280**, 17579–17583
33. Brown, G.R., Benyon, S.L., Kirk, C.J., Wictome, M., East, J.M., Lee, A.G. & Michelangeli, F. (1994) Characterisation of a novel Ca^{2+} pump inhibitor (bis-phenol) and its effects on intracellular Ca^{2+} mobilization. *Biochim. Biophys. Acta* **1195**, 252–258
34. Inesi, G., Lewis, D., Ma, H., Prasad, A. & Toyoshima, C. (2006) Concerted conformational effects of Ca^{2+} and ATP are required for activation of sequential reactions in the Ca^{2+} ATPase (SERCA) catalytic cycle. *Biochemistry* **45**, 13769–13778
35. Bilmen, J.G., Khan, S.Z., Javed, M.H. & Michelangeli, F. (2001) Inhibition of the SERCA Ca^{2+} pumps by curcumin: curcumin putatively stabilizes the interaction between the nucleotide-binding and phosphorylation domains in the absence of ATP. *Eur. J. Biochem* **268**, 6318–6327
36. Bar-Sela, G., Epelbaum, R. & Schaffer, M. (2010) Curcumin as an anti-cancer agent: review of the gap between basic and clinical applications. *Curr. Med. Chem.* **17**, 190–197
37. Toyoshima, C., Nakasako, M., Nomura, H. & Ogawa, H. (2000) Crystal structure of the calcium pump of sarcoplasmic reticulum at 2.6 Å resolution. *Nature* **405**, 647–655
38. Erlund, I., Silaste, M.L., Alfthan, G., Rantala, M., Kesäniemi, Y.A. & Aro, A. (2002) Plasma concentrations of the flavonoids hesperetin, naringenin and quercetin in human subjects following their habitual diets, and diets high or low in fruit and vegetables. *Eur. J. Clin. Nutr.* **56**, 891–898
39. Lin, Y.T., Yang, J.S., Lin, H.J., Tan, T.W., Tang, N.Y., Chaing, J.H., Chang, Y.H., Lu, H.F. & Chung, J.G. (2007) Baicalein induces apoptosis in SCC-4 human tongue cancer cells via a Ca^{2+}-dependent mitochondrial pathway. *Vivo* **21**, 1053–1058, In
40. Ogunbayo, O.A., Harris, R.M., Waring, R.H., Kirk, C.J. & Michelangeli, F. (2008) Inhibition of the sarcoplasmic/endoplasmic reticulum Ca^{2+}-ATPase by flavonoids: a quantitative structure–activity relationship study. *IUBMB Life* **60**, 853–858
41. Bilmen, J.G. & Michelangeli, F. (2002) Inhibition of the type I inositol 1,4,5-trisphosphate receptor by 2-aminoethoxydiphenylborate. *Cell. Signalling* **14**, 955–960
42. Gregory, R.B., Rychkov, G. & Barritt, G.J. (2001) Evidence that 2-aminoethyl diphenylborate is a novel inhibitor of store-operated Ca^{2+} channels in liver cells, and acts through a mechanism which does not involve inositol trisphosphate receptors. *Biochem. J.* **354**, 285–290
43. Bilmen, J.G., Wootton, L.L., Godfrey, R.E., Smart, O.S. & Michelangeli, F. (2002) Inhibition of SERCA Ca^{2+} pumps by 2-aminoethoxydiphenyl borate (2-APB): 2-APB reduces both Ca^{2+} binding and phosphoryl transfer from ATP, by interfering with the pathway leading to the Ca^{2+}-binding sites. *Eur. J. Biochem.* **269**, 3678–3687
44. Zhang, Z., Lewis, D., Sumbilla, C., Inesi, G. & Toyoshima, C. (2001) The role of the M6–M7 loop (L67) in stabilization of the phosphorylation and Ca^{2+} binding domains of the sarcoplasmic reticulum Ca^{2+}-ATPase (SERCA). *J. Biol. Chem.* **276**, 15232–15239
45. Menguy, T., Corre, F., Bouneau, L., Deschamps, S., Møller, J.V., Champeil, P., le Maire, M. & Falson, P. (1998) The cytoplasmic loop located between transmembrane segments 6 and 7 controls activation by Ca^{2+} of sarcoplasmic reticulum Ca^{2+}-ATPase. *J. Biol. Chem.* **273**, 20134–20143

46. Lee, A.G. & East, J.M. (2001) What the structure of a calcium pump tells us about its mechanism. *Biochem. J.* **356**, 665–683
47. Michelangeli, F., Orlowski, S., Champeil, P., Grimes, E.A., East, J.M. & Lee, A.G. (1990) Effects of phospholipids on binding of calcium to (Ca^{2+}-Mg^{2+})-ATPase. *Biochemistry* **29**, 8307–8312
48. Brini, M. & Carafoli, E. (2009) Calcium pumps in health and disease. *Physiol. Rev.* **89**, 1341–1378
49. Clark, J.H., Kinnear, N.P., Kalujnaia, S., Cramb, G., Fleischer, S., Jeyakumar, L.H., Wuytack, F. & Evans, A.M. (2010) Identification of functionally segregated sarcoplasmic reticulum calcium stores in pulmonary arterial smooth muscle. *J. Biol. Chem.* **285**, 13542–13549
50. Szewczyk, M.M., Pande, J. & Grover, A.K. (2008) Caloxins: a novel class of selective plasma membrane Ca^{2+} pump inhibitors obtained using biotechnology. *Pflügers Arch.* **456**, 255–266
51. Brouland, J.P., Gélébart, P., Kovàcs, T., Enouf, J., Grossmann, J. & Papp, B. (2005) The loss of sarco/endoplasmic reticulum calcium transport ATPase 3 expression is an early event during the multistep process of colon carcinogenesis. *Am. J. Pathol.* **167**, 233–242
52. Korosec, B., Glavac, D., Volavsek, M. & Ravnik-Glavac, M. (2008) Alterations in genes encoding sarcoplasmic-endoplasmic reticulum Ca^{2+} pumps in association with head and neck squamous cell carcinoma. *Cancer Genet. Cytogenet.* **181**, 112–118
53. Christensen, S.B., Skytte, D.M., Denmeade, S.R., Dionne, C., Møller, J.V., Nissen, P. & Isaacs, J.T. (2009) A Trojan horse in drug development: targeting of thapsigargins towards prostate cancer cells. *Anticancer Agents Med. Chem.* **9**, 276–294
54. Al-Mousa, F. & Michelangeli, F. (2009) Commonly used ryanodine receptor activator, 4-chloro-*m*-cresol (4CmC), is also an inhibitor of SERCA Ca^{2+} pumps. *Pharmacol. Rep.* **61**, 838–842
55. Michelangeli, F., Di Virgilio, F., Villa, A., Podini, P., Meldolesi, J. & Pozzan, T. (1991) Identification, kinetic properties and intracellular localization of the (Ca^{2+}-Mg^{2+})-ATPase from the intracellular stores of chicken cerebellum. *Biochem. J.* **275**, 555–561
56. Bilmen, J.G., Wootton, L.L. & Michelangeli, F. (2002) The mechanism of inhibition of the sarco/endoplasmic reticulum Ca^{2+} ATPase by paxilline. *Arch. Biochem. Biophys.* **406**, 55–64
57. Longland, C.L., Mezna, M. & Michelangeli, F. (1999) The mechanism of inhibition of the Ca^{2+}-ATPase by mastoparan: mastoparan abolishes cooperative Ca^{2+} binding. *J. Biol. Chem.* **274**, 14799–14805
58. Longland, C.L., Mezna, M., Langel, U., Hällbrink, M., Soomets, U., Wheatley, M., Michelangeli, F. & Howl, J. (1998) Biochemical mechanisms of calcium mobilisation induced by mastoparan and chimeric hormone–mastoparan constructs. *Cell Calcium* **24**, 27–34
59. Bilmen, J.G., Wootton, L.L. & Michelangeli, F. (2002) The inhibition of the sarcoplasmic/endoplasmic reticulum Ca^{2+}-ATPase by macrocyclic lactones and cyclosporin A. *Biochem. J.* **366**, 255–263
60. Khan, S.Z., Longland, C.L. & Michelangeli, F. (2000) The effects of phenothiazines and other calmodulin antagonists on the sarcoplasmic and endoplasmic reticulum Ca^{2+} pumps. *Biochem. Pharmacol.* **60**, 1797–1806
61. Hermanson, D., Addo, S.N., Bajer, A.A., Marchant, J.S., Das, S.G., Srinivasan, B., Al-Mousa, F., Michelangeli, F., Thomas, D.D., Lebien, T.W. & Xing, C. (2009) Dual mechanisms of sHA 14–1 in inducing cell death through endoplasmic reticulum and mitochondria. Mol. *Pharmacol.* **76**, 667–678

Biochem. Soc. Symp. 78
Citation reference: Biochem. Soc. Trans. (2011) **39**, 799–806.

12

The mitochondrial-encoded subunits of respiratory complex I (NADH:ubiquinone oxidoreductase): identifying residues important in mechanism and disease

Hannah R. Bridges, James A. Birrell and Judy Hirst[1]

Medical Research Council Mitochondrial Biology Unit, Wellcome Trust/MRC Building, Hills Road, Cambridge CB2 0XY, U.K.

Abstract

Complex I (NADH:ubiquinone oxidoreductase) is crucial to respiration in many aerobic organisms. The hydrophilic domain of complex I, containing nine or more redox cofactors, and comprising seven conserved core subunits, protrudes into the mitochondrial matrix or bacterial cytoplasm. The α-helical membrane-bound hydrophobic domain contains a further seven core subunits that are mitochondrial-encoded in eukaryotes and named the ND subunits (ND1–ND6 and ND4L). Complex I couples the oxidation of NADH in the hydrophilic domain to ubiquinone reduction and proton translocation

[1] *To whom correspondence should be addressed (email jh@mrc-mbu.cam.ac.uk).*

in the hydrophobic domain. Although the mechanisms of NADH oxidation and intramolecular electron transfer are increasingly well understood, the mechanisms of ubiquinone reduction and proton translocation remain only poorly defined. Recently, an α-helical model of the hydrophobic domain of bacterial complex I [Efremov, Baradaran and Sazanov (2010) Nature **465**, 441–447] revealed how the 63 transmembrane helices of the seven core subunits are arranged, and thus laid a foundation for the interpretation of functional data and the formulation of mechanistic proposals. In the present paper, we aim to correlate information from sequence analyses, site-directed mutagenesis studies and mutations that have been linked to human diseases, with information from the recent structural model. Thus we aim to identify and discuss residues in the ND subunits of mammalian complex I which are important in catalysis and for maintaining the enzyme's structural and functional integrity.

The membrane-bound domain of complex I

Complex I (NADH:ubiquinone oxidoreductase) is a complicated multisubunit enzyme found in the mitochondrial inner membrane of many eukaryotes and in the cytoplasmic membrane of many aerobic bacteria. It couples the transfer of two electrons, from NADH to ubiquinone, to the translocation of four protons across the membrane. NADH is oxidized by an FMN in the hydrophilic domain of the enzyme (that protrudes into the mitochondrial matrix or bacterial cytoplasm), then the electrons are transferred along a series of iron–sulfur clusters to the highly hydrophobic ubiquinone acceptor, bound in, or close to, the membrane domain [1–3]. Structural models for the hydrophilic domain of complex I from *Thermus thermophilus*, containing the flavin and FeS clusters, have been determined [4], and the electron-transfer reactions within this domain are increasingly well understood [1]; conversely, the mechanisms of ubiquinone reduction and proton translocation remain only poorly understood.

Complex I from *Bos taurus* is closely related to the human enzyme, and is known to contain 45 different subunits [5]. A total of 14 core subunits are conserved in all complexes I; the additional supernumerary subunits are generally specific to eukaryotes and vary between species [6]. There are seven core subunits in the hydrophilic domain, and seven in the hydrophobic domain. The hydrophobic core subunits (named ND1–ND6 and 4L) are encoded by the mitochondrial genome in mammals [7] and most eukaryotes (*Chlamydomonas reinhardtii* is an exception that encodes ND3 and ND4L in its nuclear genome [8]). Recently, new structural information about the membrane domains of the complexes I from *Escherichia coli*, *T. thermophilus* [2] and *Yarrowia lipolytica* [3] (see Figure 1) has been described, forming a foundation for the interpretation of functional data and the formulation of mechanistic proposals for ubiquinone reduction and proton translocation. The present review focuses on the seven mitochondrial-encoded subunits in mammalian complex I, aiming to relate

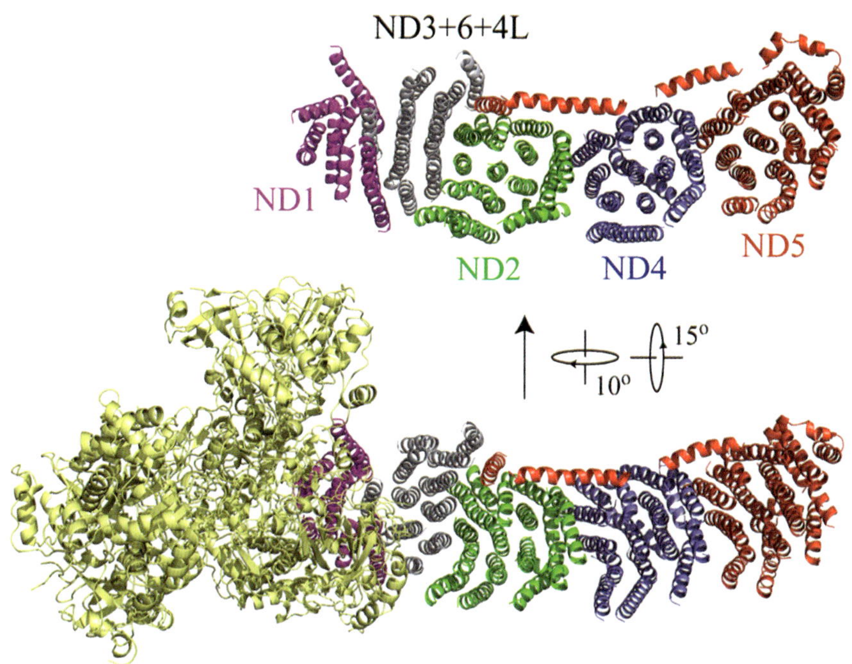

Figure 1. The arrangement of transmembrane helices in the membrane domain of complex I from *T. thermophilus*
The subunits are ND1 (magenta, eight helices), ND2 (green, 14 helices), ND4 (blue, 14 helices), ND5 (red, 16 transmembrane helices plus lateral helix), ND3+ND6+ND4L (grey, 11 helices). In the lower panel, the hydrophilic domain (yellow) is protruding from the page and the molecule is viewed from above the membrane, showing that many of the helices are tilted. In the upper panel, the molecule has been rotated to display more clearly the structural similarity of ND2, ND4 and ND5. Taken from PDB code 3I9V [2].

information from sequence analyses, site-directed mutagenesis and disease-linked mutations to recent structural data.

Correlations between structurally observed transmembrane helices and the protein sequences

Figure 1 shows the currently available α-helical model of the membrane domain of complex I from *T. thermophilus* [2]. There are two striking features: the structural symmetry between three groups of 14 helices that comprise the distal section of the domain, and the lateral (transverse) helix that runs in the plane of the membrane, suggestive of a mechanical linkage. The three 14-helix units are assigned to subunits ND2, ND4 and ND5, which are all related to one another, and to proteins from the Mrp Na^+/H^+ antiporter family [9]. Two 'broken' helices, indicative of proton-translocation loci, were observed in each of the three subunits; in each subunit, one of them contacts the lateral helix. Subunit

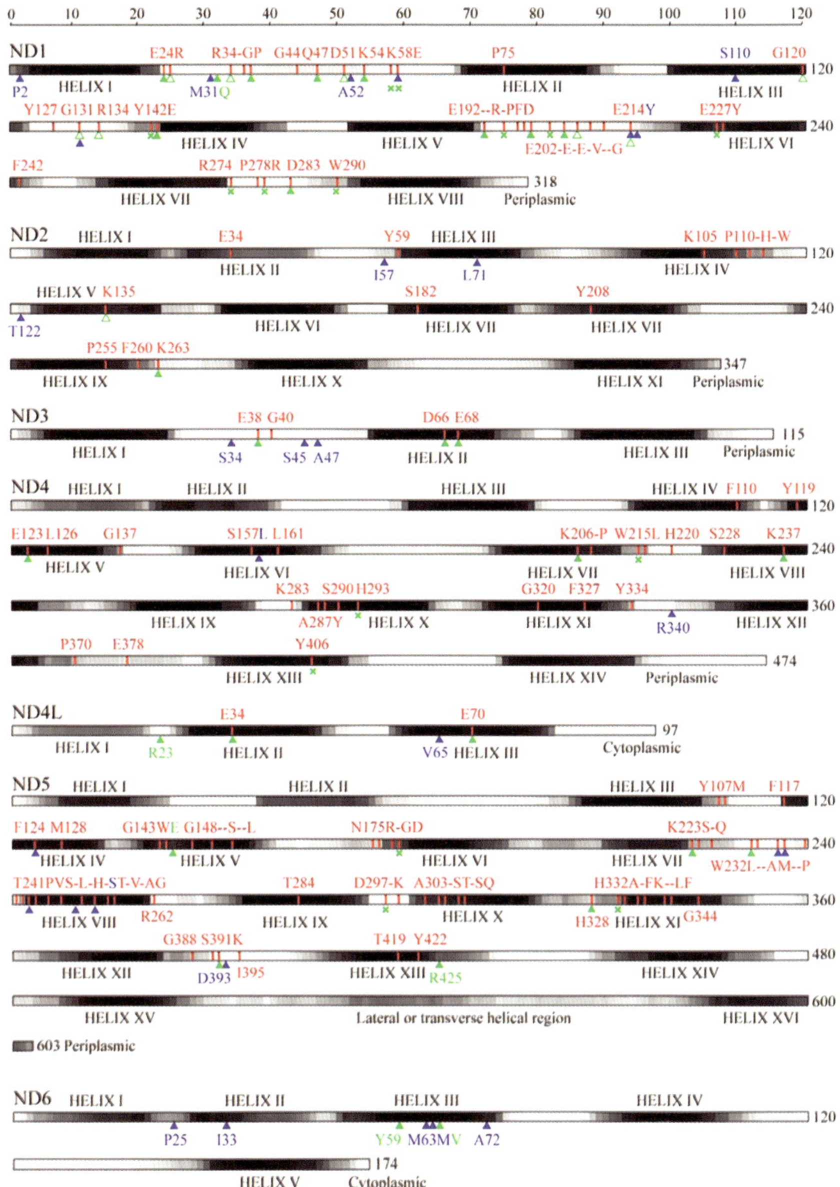

Figure 2. Summary of information about the sequences of the human ND subunits from sequence alignments

The shaded bars represent the transmembrane helical structures of each subunit, constructed as described in the text: black, helix predicted in all sequences aligned; white, helix predicted in no sequences. Residues marked with red lines are considered to be conserved: they are identical in at least 38 of the 40 species aligned. For consecutive residues, the number of the first residue is given, gaps in the residues labelled are indicated by dashes. Residues marked in green have been mutated in bacterial model systems. Closed green triangles indicate residues that, when mutated, lead to decreases in the catalytic activity to <50% of the wild-type value. Open green triangles indicate residues that, when mutated, lead to disruption of the enzyme structure or assembly. Conserved

ND5 is the terminal subunit; it has a C-terminal sequence extension, relative to ND2 and ND4, consistent with two additional helices and the lateral helix. Subunits ND4 and ND5 form the core of subcomplex Iβ from *B. taurus* complex I [6], and can be resolved together from *E. coli* complex I [10], so subunit ND4 is the central subunit, and ND2 is closest to the hydrophilic arm. The eight helices that comprise subunit ND1 were identified by comparing the data from *T. thermophilus* complex I (Figure 1) with data from the membrane domain of *E. coli* complex I lacking ND1 [2]. The remaining 11 helices in Figure 1 cannot be assigned to specific subunits at present; they are formed together by subunits ND3, ND6 and ND4L. The hydrophilic domain interacts predominantly with ND1, but also with the ND3, ND6 and ND4L bundle; the binding site for the quinone headgroup is probably located close to the interfacial region between the domains.

Figure 2 summarizes information from our analyses of the sequences of the seven ND subunits, based on the sequences of the human subunits. An alignment was generated for each subunit, using the sequences of 40 organisms chosen to represent all parts of the complex I-encoding phylogenetic tree. Then, the positions of the transmembrane helices in each individual sequence were predicted, and the consensuses shown in Figure 2 were generated by counting the number of sequences for which a helix was predicted at each position (see [11] for details). There is a clear consensus for eight helices in ND1, consistent with previous proposals [12–14] and with the structural analysis described above; the long loop region between helices I and II is thought to be on the cytoplasmic (matrix) side [13]. A total of 11 helices are predicted clearly for ND2; the lack of the three N-terminal helices in bilaterian complexes I (including mammals), relative to the 14-helix subunit in fungi, bacteria and lower eukaryotes (consistent with the structural data and as proposed previously [15]), has been discussed previously [11]. The 14-helix homologue is thought to begin on the periplasmic side [15], so, assuming that the orientation is conserved, the 11-helix homologue begins on the cytoplasmic (matrix) side. The truncation may represent a loss or change in function for ND2 in the bilaterian complex, or it may suggest that the three N-terminal helices of ND2 in lower organisms (and, by extension, the corresponding helices in ND4 and ND5) are not essential for function. Alternatively, the truncation of ND2 may have facilitated its recruitment to another role in the cell; ND2 has been shown to form a complex with Src kinase and the NDMA (*N*-methyl-D-aspartate) receptor at synaptic membranes in the brain, a function that requires its export from the mitochondrion [16]. There are 14 helices predicted for ND4, consistent with the structural data; previously, 14 helices were proposed, starting on the periplasmic side, but it

residues which were mutated, but did not have a significant effect on catalysis, are marked with green crosses. Residues marked with a blue triangle are substitution positions associated with complex I-related diseases (see Table 1).

was not clear whether helices XI and XII were transmembrane helices or not [9,17]. A total of 16 transmembrane helices can be identified for ND5, again consistent with the structural data and previous proposals, and starting on the periplasmic side [9,18]; the lateral helix between transmembrane helices XV and XVI is an unusual feature that is not expected to be analysed meaningfully by transmembrane helix predictions. The three remaining subunits, ND3, ND4L and ND6, are the smallest subunits, and they are predicted to have three, three and five transmembrane helices respectively, consistent with previous proposals [19–25] and with the 11 helices observed in the structural analysis (the predicted topologies are marked on Figure 2).

Sequence conservation in the ND subunits

The ND subunits from very few species have been characterized directly (they are difficult to identify and characterize by MS owing to their hydrophobicity and lack of protease cleavage sites [26]), so the majority of the sequences compared in Figure 2 are translations of genome sequences: they rely on identification of the relevant stretches of sequence by homology searches, use of the appropriate translation code and (particularly in fungi and plants) the correct identification of introns [27]. The sequence alignments reveal a general trend for shorter sequences in the higher eukaryotes; the insertions in the sequences of the lower organisms are not included in Figure 2, but they lend flexibility to the alignments (aiding 'identification' of conserved residues, particularly in loop regions). With this in mind, the conserved residues marked on Figure 2 are those that are identical in at least 38 of the 40 sequences analysed.

Subunit ND1 contains 37 conserved residues (11.6%), located predominantly in the loop regions on the cytoplasmic (matrix) side; these loops are probably important in forming contacts with the hydrophilic domain. There are conserved residues in the interfacial regions also, but relatively few in the helical regions; of these, only one is charged (Glu^{227} in helix VI). There are only 12 conserved residues (3.5 %) in subunit ND2; most notable are Glu^{34} in helix II and Lys^{105} in helix IV. Subunits ND4 and ND5 contain 24 and 54 conserved residues respectively (5.1 % and 9.0 %); the lack of conservation in the first three helices (homologous with the three helices absent from ND2) is notable, and consistent with the suggestion that these helices are not important functionally. In contrast with ND1, most of the conserved residues in ND2, ND4 and ND5 fall within helices, supporting their proposed roles in proton translocation. Two conserved charged residues in ND4 and ND5, Glu^{123} and Glu^{145} in helix V (Glu^{145} is not marked as conserved on Figure 2, but there is always an acidic residue (glutamate or aspartate) at this position, and Lys^{206} and Lys^{223} in helix VII, are equivalent to the two conserved residues in helices II and IV in ND2, and they are conserved also in MrpA and MrpD of the Mrp Na^+/H^+ antiporters [9]. It is possible that these residues produce the two broken helices observed in the structural analysis. There are further conserved charged residues in the helical regions in ND2 (Lys^{135} in helix V), ND4 (Lys^{237} in helix VIII, analogous to Lys^{135} in ND2) and ND5 (Lys^{336} in helix XI). Interestingly, although there are several highly conserved regions in ND5 (notably helices V, VIII, X and XI, and the loop between VII and VIII), there is no apparent conservation within the lateral helix, or in helices XV or XVI. Subunits ND3 and ND4L are poorly conserved [four and two conserved residues (3.5 % and 2.1 %) respectively], but there are two charged conserved residues in helices in each subunit, suggestive of possible roles in proton translocation: Asp^{66} and Glu^{68} are conserved in helix II in ND3 (plus Glu^{38} in the periplasmic loop between helices I and II) and Glu^{34} and Glu^{70} are conserved in helices II and III in ND4L. There are no conserved residues in subunit ND6.

Studies of site-directed mutations in bacterial model systems

The difficulty of genetically manipulating the mitochondrial DNA in eukaryotic systems means that mutagenesis studies on the ND subunits are usually carried out in bacterial model systems: often in *E. coli*, but also in *Paracoccus denitrificans* and *Rhodobacter capsulatus*. Figure 2 indicates those residues for which at least one mutation that decreases the catalytic activity to less than 50 % of the wild-type value has been studied. Here, we refer to the NADH:quinone (or NADH:O_2) oxidoreduction rate, normalized for the amount of complex I present, as the 'catalytic activity' or 'activity'. Typically, the amount of complex I is quantified via the rate of NADH oxidation (catalysed in the hydrophilic domain) induced by a hydrophilic artificial electron acceptor such as $K_3[Fe(CN)_6]$. Only residues which are the same in humans and in the model system are included in Figure 2, and conserved residues for which all the mutants studied so far have relative catalytic activities above 50 % are marked with crosses. In the following sections, residue numbers are given for the human sequences, with the bacterial positions in parentheses for cross-referencing.

Several studies have described mutations of residues in the bacterial homologues of subunit ND1 [14,28,29]. Recently, Sinha and coworkers identified a set of residues {Arg^{25} (37), Arg^{34} (46), Asp^{51} (63), Gly^{120} (134), Gly^{131} (145), Arg^{134} (148), Glu^{206} (220) and Glu214 (228) [14]} that, when mutated, disrupt the structural integrity of the enzyme: the enzymes exhibit low NADH:ferricyanide oxidoreduction activities and are not assembled fully. These residues (marked by open green triangles in Figure 2) are located in the cytoplasmic (matrix) loops of ND1, supporting their proposed importance for the connection with the hydrophilic domain. Further mutations in ND1 that compromise catalysis are shown on Figure 2; they are also located in the cytoplasmic loops, and it has been suggested that some of them affect quinone binding [28]. No significant effect on catalysis was observed from mutating Glu^{227} in helix VI [14].

Mutations of the two charged residues that are conserved throughout ND2 (Glu^{34} in helix II, and Lys^{105} in helix IV), ND4 and ND5 (Glu^{123} and Glu^{145} in helix V, and Lys^{206} and Lys^{223} in helix VII) have been studied. In ND2 [15], there was little apparent effect from mutating Glu^{34} (133), and the origin of the effect of mutating Lys^{105} (217) is unclear because the complex I was not quantified. In ND4, the E123A (144) and E123Q mutations (but not the E123D mutation), and the K206A (234) and K206R mutations, essentially abolished the activity [30,31]. In ND5, the E145A (144) and E145Q variants were ~20 % active, and the K223A (229), K223R and K223E variants decreased the activity to 10–25 % [32]. Thus these studies are generally consistent with important roles for these residues in catalysis, although an unambiguous picture has yet to emerge. For the conserved lysine residue in ND2 and ND4 (Lys^{135} in helix V and Lys^{237} in helix VIII) the K135C (247) mutant resulted in very low levels of complex I and the K135R mutant retained ~80 % activity (the complex I was not quantified) [15]; the K237A (237) mutation decreased the activity moderately, to 30–60 % [30,31]. These results suggest a secondary role for this residue, perhaps in maintaining structural integrity. Further residues for which mutations decreased the normalized catalytic activity significantly, and conserved residues for which they did not, are indicated in Figure 2 [30–32]. So far, no mutations in the lateral helix of ND5 or the two helices adjacent to it have been studied.

In subunit ND3, the E38A (51), D66A (79), E68Q (81) and E68A mutations retained between 30 and 60 % activity, whereas D66N had little effect, and the double mutant D66N/E68Q abolished the activity [20]. In subunit ND4L, the E34A and E34Q (36) mutations abolished the activity, whereas the E70Q (72), E70A, R23A (25) and R23K mutations retained 25–45 % activity [22,23]. In subunit ND6, the Y59F (59), V65G (65) and V65L mutations decreased the catalytic activity to between 10 and 45 % [25,33]. The results clearly suggest an important role for Glu^{34} in ND4L, but, as with many of the mutations described in ND2, ND4 and ND5, the moderate decreases in activity associated with the other mutations are hard to interpret.

Clearly, there are several factors to consider when evaluating studies of site-directed mutations in bacterial complex I model systems.

(i) Mutations may disrupt the assembly or structural integrity of the enzyme. It is common practice to assess the amount of complex I present by measuring the NADH dehydrogenase activity catalysed by the hydrophilic domain, and by using immunodetection to determine whether the full subunit complement is present. Mutations which disrupt the structural integrity may highlight regions that form important structural contacts between domains (as proposed for ND1), but they may also arise from less pertinent effects, such as steric clashes or hydrophobic mismatches. Furthermore, the presence of a subunit in a membrane preparation does not demonstrate that it is properly folded or incorporated into the complex, an issue that can perhaps be addressed by enzyme purification.

(ii) How significant does a decrease in the catalytic activity need to be to define a catalytically 'critical' residue (rather than a 'secondary' residue that is required to, for example, maintain a structural arrangement or interaction)? Perhaps a cut-off value of 10% activity is reasonable, but then should every mutation of the given residue attain this cut-off? Of course, inappropriate replacements may introduce negative effects on catalysis, rather than remove positive ones. For example, the E123A (144) and E123Q mutations in ND4 essentially abolished the activity, suggesting that the carboxylate functionality is critical [30,31]. On the other hand, the E123D variant retained full activity, and moving the carboxylate residue up or down the helix partially restored the activity [17]. Thus there is considerable flexibility in the position of the carboxylate headgroup, rather suggesting that Glu^{123} is unlikely to function as a precisely positioned gating residue in proton transfer.

(iii) Equivalent mutations may behave differently in different systems, because of interactions with non-conserved residues. The nuclear-encoded supernumerary MWFE subunit, which is not well conserved and that has no primary role in catalysis, provides a pertinent example. When several residues were mutated to their human counterparts in CHO (Chinese-hamster ovary) cells, the enzyme activity was reduced; furthermore, a single, conservative point mutation (arginine to lysine) produced a catalytically inactive enzyme [34]. These results illustrate both how the same residues can exert very different effects in even closely related systems, dictated by the 'background' of the nuclear and/or mitochondrial protein sequences of the other subunits, and how an apparently innocuous change can produce such significant (and potentially misleading) results.

(iv) There is a clear requirement for methods to distinguish which step in the catalytic mechanism of complex I is affected by a mutation [1]. The reactions of the flavin can be studied in isolation, and mutations which affect substrate binding may be tentatively assigned by identifying changes in K_m, but otherwise (unless the mutation uncouples the enzyme), all the steps are tightly coupled: inhibiting any one of them inhibits them all. In particular, measurements of proton translocation, often done by using dyes such as oxanol and Acridine Orange, are affected by many different experimental variables and (in our opinion) are not yet sufficiently well developed or understood to provide definitive information.

Mutations implicated as the causes of human diseases

Mutations in the mitochondrial-encoded subunits of complex I are increasingly implicated in a variety of diseases, including neurodegenerative diseases, optic neuropathies and diabetes [35]. In some cases, for example the 3460 mutation in ND1 and the 11778 mutation in ND4 which cause Leber's hereditary optic neuropathy, the link between mutation and disease is well established, but in many other cases, distinguishing a cause-and-effect relationship from a polymorphism is difficult. Many patients present multiple deviations from the reference mitochondrial genome sequence, and the relationship is complicated further by variable levels of heteroplasmy. We applied the scoring system of Turnbull and co-workers [35] to the list of disease-related mutations in the MitoMap online database [36]. We discuss mutations which scored above 20 (out of 40) in this evaluation, which uses biochemical deficiencies, number of independent reports, heteroplasmy, matrilineal variant segregation and conservation as its criteria. Turnbull and co-workers analysed the 50 mutations listed in MitoMap in 2006 and found that 16 were very likely to be pathogenic, three probably, ten possibly and the rest almost certainly not [35]: in the present review, we consider these 16 mutations, plus further mutations identified from recent literature. The mutations discussed are summarized in Table 1 and indicated on Figure 2.

First, the majority of the disease-associated residues in Table 1 are not highly conserved; those that are conserved are in ND1 and ND5 only. Thus, although conservation is often taken as an indicator of relevance, it is certainly not a strong determinant. Indeed, mutations of critical catalytic residues that abolish activity may be less likely to present as diseases, and mutations which compromise the activity may be tolerated more easily when the mutation is heteroplasmic. Secondly, the disease-associated residues are distributed throughout the helix and loop regions; in ND1 and ND3 the majority are in the loop regions, whereas in ND5 and ND6, the majority are in the helical regions. Thirdly, equivalent mutations have been made in bacterial model systems for only nine of the mutations listed in Table 1. For ND1-A52T, alanine to methionine and threonine mutations decreased the activity only slightly (80–90 %), but they may affect quinone binding [28]; for ND1-E59K, a glutamine to alanine mutation resulted in a decreased level of complex I, and a moderately decreased activity (~60 %) [14]; for ND1-G131S, a glycine to valine mutation had a significant effect (decreased enzyme content and very low activity), but a glycine to alanine substitution had less effect (~50 % activity) [14]; for ND1-E214A, the direct substitution caused a significant decrease in both enzyme content and catalytic activity, with similarly large effects from other substitutions [14,29]; the direct ND1-Y215H mutation had no apparent effect on the enzyme activity [29]. For ND4, neither the R340H or R340A substitutions had significant effects on catalysis in *R. capsulatus* (although *E. coli* R340H was only ~70 % active), but cell growth and metabolism were affected [30,37]. For ND5-D393N/G, substitutions with asparagine, glutamate or alanine produced variable, but limited, changes in enzyme activity (>80 %) [32]. For ND6-A72V, the methionine to valine, alanine, cysteine and isoleucine substitutions had variable effects (the most significant decrease was for M72V, ~40 %), but the starting residue in the bacterial system is different [33]; for ND6-M64V, the methionine to isoleucine or valine substitutions had no, or limited, effects [25,33]. Thus in only one case (ND1-E214A) is the direct disease-linked mutation in the bacterial system known to cause a severe defect; for ND1-Gly^{131}, severe effects were observed, but for a different substitution. It is clear that the value of these studies for understanding the mutations in human complex I is compromised unless the

Table 1. Mutations in the mitochondrial-encoded subunits of complex I that are associated with human disease phenotypes

The human nucleotide and amino acid substitutions for the disease-related mutations are given, along with an example of a disease associated with the mutation. Only mutations which scored at least 20 out of 40 using the evaluation method of Mitchell et al. [35], are included, and one reference for each mutation was chosen from the MitoMap database [36]. The Table is not a complete survey of all disease-associated mutations, but provides a framework for the discussion of point mutations which are likely to have an effect on the enzyme. LHON, Leber's hereditary optic neuropathy; MELAS, mitochondrial myopathy, encephalopathy, lactic acidosis and stroke.

Subunit	DNA	Protein	Medical condition	Reference
ND1	C3310T	P2S	Type 2 diabetes and cardiomyopathy	[38]
	T3398C	M31T	MELAS, cardiomyopathy and Type 1 diabetes	[39]
	G3460A	A52T	LHON	[35]
	G3481A	E59K	MELAS	[40]
	G3635A	S110N	LHON	[35]
	G3697A	G131S	MELAS and Leigh	[35]
	G3946A	E214K	MELAS	[35]
	T3949C	Y215H	MELAS	[35]
ND2	C4640A	I57M	LHON	[35]
	T4681C	L71P	Leigh	[41]
	A4833G	T122A	Type 2 diabetes	[42]
ND3	T10158C	S34P	Leigh	[35]
	T10191C	S45P	Leigh	[35]
	G10197A	A47T	Leigh and dystonia	[43]
ND4	T11232C	L158P	Progressive external ophthalmoplegia	[35]
	C11777A	R340S	Leigh	[35]
	G11778A	R340H	LHON	[35]
	G11832A	W358ter	Exercise intolerance	[35]
ND4L	T10663C	V65A	LHON	[35]
ND5	T12706C	F124L	Leigh	[35]
	G13042A	A236T	MELAS and Leigh	[44]
	A13045C	M237L	MELAS, LHON and Leigh	[35]
	G13063A	V243I	MELAS	[40]
	A13084T	S250C	MELAS and Leigh	[35]
	T13094C	V253A	Ataxia and progressive external ophthalmoplegia	[45]
	G13513A	D393N	MELAS and Leigh	[35]
	A13514G	D393G	MELAS	[35]
ND6	G14600A	P25L	Leigh	[40]
	T14577C	I33V	Type 2 diabetes	[46]
	T14487C	M63V	Leigh and cardiomyopathy	[40]
	T14484C	M64V	LHON	[35]
	G14459A	A72V	LHON, dystonia and Leigh	[35]

same starting and ending residues are applied. Even then, the backgrounds for the mutations differ significantly between the human enzyme and the bacterial homologues, so that observed effects, which may be both very subtle and depend on interactions with other nuclear and mitochondrial-encoded proteins, should not be overinterpreted. It is also important to note that effects that may be considered insignificant in mechanistic studies (for example, retention of 80% of the activity) may be very significant in a physiological context. Finally, simple catalytic rate measurements may not reveal important defects (for example, an enzyme may be compromised only in its ability to work against a substantial protonmotive force, or under certain physiological conditions), underlining the requirement for developing improved analytical and mechanistic approaches.

Funding

Research in our laboratory is supported by the Medical Research Council.

References

1. Hirst, J. (2010) Towards the molecular mechanism of respiratory complex I. *Biochem. J.* **425**, 327–339
2. Efremov, R.G., Baradaran, R. & Sazanov, L.A. (2010) The architecture of respiratory complex I. *Nature* **465**, 441–447
3. Hunte, C., Zickermann, V. & Brandt, U. (2010) Functional modules and structural basis of conformational coupling in mitochondrial complex I. *Science* **329**, 448–451
4. Sazanov, L.A. & Hinchliffe, P. (2006) Structure of the hydrophilic domain of respiratory complex I from *Thermus thermophilus*. *Science* **311**, 1430–1436
5. Carroll, J., Fearnley, I.M., Skehel, J.M., Shannon, R.J., Hirst, J. & Walker, J.E. (2006) Bovine complex I is a complex of forty-five different subunits. *J. Biol. Chem.* **281**, 32724–32727
6. Hirst, J., Carroll, J., Fearnley, I.M., Shannon, R.J. & Walker, J.E. (2003) The nuclear encoded subunits of complex I from bovine heart mitochondria. *Biochim. Biophys. Acta* **1604**, 135–150
7. Chomyn, A., Cleeter, M.W.J., Ragan, C.I., Riley, M., Doolittle, R.F. & Attardi, G. (1986) URF6, last unidentified reading frame of human mtDNA, codes for an NADH dehydrogenase subunit. *Science* **234**, 614–618
8. Cardol, P., Lapaille, M., Minet, P., Franck, F., Matagne, R. & Remacle, C. (2006) ND3 and ND4L subunits of mitochondrial complex I, both nucleus encoded in *Chlamydomonas reinhardtii*, are required for activity and assembly of the enzyme. *Eukaryotic Cell* **5**, 1460–1467
9. Mathiesen, C. & Hägerhäll, C. (2002) Transmembrane topology of the NuoL, M and N subunits of NADH:ubiquinone oxidoreductase and their homologues among membrane-bound hydrogenases and *bona fide* antiporters. *Biochim. Biophys. Acta* **1556**, 121–132
10. Baranova, E.A., Morgan, D.J. & Sazanov, L.A. (2007) Single particle analysis confirms distal location of subunits NuoL and NuoM in *Escherichia coli* complex I. *J. Struct. Biol.* **159**, 238–242
11. Birrell, J.A. & Hirst, J. (2010) Truncation of subunit ND2 disrupts the threefold symmetry of the antiporter-like subunits in complex I from higher metazoans. *FEBS Lett.* **584**, 4247–4252
12. Kurki, S., Zickermann, V., Kervinen, M., Hassinen, I. & Finel, M. (2000) Mutagenesis of three conserved Glu residues in a bacterial homologue of the ND1 subunit of complex I affects ubiquinone reduction kinetics but not inhibition by dicyclohexylcarbodiimide. *Biochemistry* **39**, 13496–13502

13. Roth, R. & Hägerhäll, C. (2001) Transmembrane orientation and topology of the NADH:quinone oxidoreductase putative quinone binding subunit NuoH. *Biochim. Biophys. Acta* **1504**, 352–362
14. Sinha, P.K., Torres-Bacete, J., Nakamaru-Ogiso, E., Castro-Guerrero, N., Matsuno-Yagi, A. & Yagi, T. (2009) Critical roles of subunit NuoH (ND1) in the assembly of peripheral subunits with the membrane domain of *Escherichia coli* NDH-1. *J. Biol. Chem.* **284**, 9814–9823
15. Amarneh, B. & Vik, S.B. (2003) Mutagenesis of subunit N of the *Escherichia coli* complex I. Identification of the initiation codon and the sensitivity of mutants to decylubiquinone. *Biochemistry* **42**, 4800–4808
16. Gingrich, J.R., Pelkey, K.A., Fam, S.R., Huang, Y., Petralia, R.S., Wenthold, R.J. & Salter, M.W. (2004) Unique domain anchoring of Src to synaptic NMDA receptors via the mitochondrial protein NADH dehydrogenase subunit 2. *Proc. Natl. Acad. Sci. U.S.A.* **101**, 6237–6242
17. Torres-Bacete, J., Sinha, P.K., Castro-Guerrero, N., Matsuno-Yagi, A. & Yagi, T. (2009) Features of subunit NuoM(ND4) subunit in *Escherichia coli* NDH-1. *J. Biol. Chem.* **284**, 33062–33069
18. Steuber, J. (2003) The C-terminally truncated NuoL subunit (ND5 homologue) of the Na-dependent complex I from *Escherichia coli* transports Na^+. *J. Biol. Chem* **278**, 26817–16822
19. Di Bernardo, S., Yano, T. & Yagi, T. (2000) Exploring the membrane domain of the reduced nicotinamide adenine dinucleotide-quinone oxidoreductase of *Paracoccus denitrificans*: characterisation of the NQO7 subunit. *Biochemistry* **39**, 9411–9418
20. Kao, M.-C., Di Bernardo, S., Perego, M., Nakamaru-Ogiso, E., Matsuno-Yagi, A. & Yagi, T. (2004) Functional roles of four conserved charged residues in the membrane domain subunit NuoA of the proton-translocating NADH:Q oxidoreductase from *Escherichia coli*. *J. Biol. Chem* **279**, 32360–32366
21. Kao, M.-C., Di Bernardo, S., Matsuno-Yagi, A. & Yagi, T. (2002) Characterisation of the membrane domain Nqo11 subunit of the proton-translocating NADH-quinone oxidoreductase of *Paracoccus denitrificans*. *Biochemistry* **41**, 4377–4384
22. Kervinen, M., Pätsi, J., Finel, M. & Hassinen, I. (2004) A pair of membrane embedded acidic residues in the NuoK subunit of *Escherichia coli* NDH-1, a counterpart of the ND4L subunit of complex I, are required for high ubiquinone reductase activity. *Biochemistry* **43**, 773–781
23. Kao, M.-C., Nakamaru-Ogiso, E., Matsuno-Yagi, A. & Yagi, T. (2005) Characterisation of the membrane domain NuoK (ND4L) NADH-quinone oxidoreductase of *Escherichia coli*. *Biochemistry* **44**, 9545–9554
24. Kao, M.-C., Di Bernardo, S., Matsuno-Yagi, A. & Yagi, T. (2003) Characterisation and topology of the membrane domain Nqo10 subunit of the proton-translocating NADH-quinone oxidoreductase of *Paracoccus denitrificans*. *Biochemistry* **42**, 4534–4543
25. Kao, M.-C., Di Bernardo, S., Nakamaru-Ogiso, E., Miyoshi, H., Matsuno-Yagi, A. & Yagi, T. (2005) Characterisation of the membrane domain NuoJ (ND6) NADH-quinone oxidoreductase of *Escherichia coli* by chromosomal DNA manipulation. *Biochemistry* **44**, 3562–3571
26. Carroll, J., Altman, M.C., Fearnely, I.M. & Walker, J.E. (2007) Identification of membrane proteins by tandem mass spectrometry of protein ions. *Proc. Natl. Acad. Sci. U.S.A.* **104**, 14330–14335
27. Lang, B.F., Laforest, M.-J. & Burger, G. (2007) Mitochondrial introns: a critical view. *Trends Genet.* **23**, 119–125
28. Zickermann, V., Barquera, B., Wikström, M. & Finel, M. (1998) Analysis of the pathogenic human mitochondrial mutation ND1/3460, and mutations of strictly conserved residues in its vicinity, using the bacterium *Paracoccus denitrificans*. *Biochemistry* **37**, 11792–11796
29. Kervinen, M., Hinttala, R., Helander, H.M., Kurki, S., Uusimaa, J., Finel, M., Majamaa, K. & Hassinen, I. (2006) The MELAS mutations 3946 and 3949 perturb the critical structure in a conserved loop of the ND1 subunit of mitochondrial complex I. *Hum. Mol. Genet.* **15**, 2543–2552
30. Torres-Bacete, J., Nakamaru-Ogiso, E., Matsuno-Yagi, A. & Yagi, T. (2007) Characterisation of the NuoM(ND4) subunit in *Escherichia coli* NDH-1. *J. Biol. Chem.* **282**, 36914–36922
31. Euro, L., Belevich, G., Verkhovsky, M.I., Wikström, M. & Verkhovskaya, M. (2008) Conserved lysine residues of the membrane subunit NuoM are involved in energy conversion by the proton-pumping NADH:ubiquinone oxidoreductase (complex I). *Biochim. Biophys. Acta* **1777**, 1166–1172

32. Nakamura-Ogiso, E., Kao, M.-C., Chen, H., Sinha, S.C., Yagi, T. & Ohnishi, T. (2010) The membrane subunit NuoL(ND5) is involved in the indirect proton pumping mechanism of *Escherichia coli* complex I. *J. Biol. Chem.* **285**, 39070–39078
33. Pätsi, J., Kervinen, M., Finel, M. & Hassinen, I.E. (2008) Leber hereditary optic neuropathy mutations in the ND6 subunit of mitochondrial complex I affect ubiquinone reduction kinetics in a bacterial model of the enzyme. *Biochem. J.* **409**, 129–137
34. Yadava, N., Potluri, P., Smith, E.N., Bisevac, A. & Scheffler, I.E. (2002) Species specific and mutant MWFE proteins. *J. Biol. Chem.* **277**, 21221–21230
35. Mitchell, A.L., Elson, J.L., Howell, N., Taylor, R.W. & Turnbull, D.M. (2006) Sequence variation in mitochondrial complex I genes: mutation or polymorphism? *J. Med. Genet.* **43**, 175–179
36. Brandon, M.C., Lott, M.T., Nguyen, K.C., Spolim, S., Navathe, S.B., Baldi, P. & Wallace, D.C. (2005) MITOMAP: a human mitochondrial genome database – 2004 update. *Nucleic Acids Res.* **33**, D611–D613
37. Lunardi, J., Darrouzet, E., Dupuis, A. & Issartel, J.-P. (1998) The *nuoM* $arg^{368}his$ mutation in NADH:ubiquinone oxidoreductase from *Rhodobacter capsulatus*: a model for the human *nd4*–11778 mtDNA mutation associated with Leber's hereditary optic neuropathy. Biochim. Biophys. *Acta* **1407**, 114–124
38. Chen, J., Hattori, Y., Nakajima, K., Eizawa, T., Ehara, T., Koyama, T., Hirai, T., Fukuda, Y., Kinoshita, M., Sugiyama, A. et al. (2006) Mitochondrial complex I activity is significantly decreased in a patient with maternally inherited type 2 diabetes mellitus and hypertrophic cardiomyopathy associated with mitochondrial DNA C3310T mutation: a cybrid study. *Diabetes Res. Clin. Pract.* **74**, 148–153
39. Jaksch, M., Hofmann, S., Kaufhold, P., Obermaier-Kusser, B., Zierz, S. & Gerbitz, K.-D. (1996) A novel combination of mitochondria l tRNA and *ND1* gene mutations in a syndrome with MELAS, cardiomyopathy, and diabetes mellitus. *Hum. Mutat.* **7**, 358–360
40. Malfatti, E., Bugiani, M., Invernizzi, F., Fischinger-Moura de Souza, C., Farina, L., Carrara, F., Lamantea, E., Antozzi, C., Confalonieri, P., Sanseverino, M.T. et al. (2007) Novel mutations of ND genes in complex I deficiency associated with mitochondrial encephalopathy. *Brain* **130**, 1894–1904
41. Ugalde, C., Hinttala, R., Timal, S., Smeets, R., Rodenburg, R.J.T., Uusimaa, J., van Heuvel, L.P., Nijtmans, L.G.J., Majamaa, K. & Smeitink, J.A.M. (2007) Mutated ND2 impairs mitochondrial complex I assembly and leads to Leigh syndrome. *Mol. Genet. Metab.* **90**, 10–14
42. Ohkubo, E., Aida, K., Chen, J., Hayashi, J.-I., Isobe, K., Tawata, M. & Onaya, T. (2000) A patient with type 2 diabetes mellitus associated with mutations in calcium sensing receptor gene and mitochondrial DNA. *Biochem. Biophys. Res. Commun.* **278**, 808–813
43. Sarzi, E., Brown, M.D., Lebon, S., Chretien, D., Munnich, A., Rotig, A. & Procaccio, V. (2007) A novel recurrent mitochondrial DNA mutation in *ND3* gene is associated with isolated complex I deficiency causing Leigh syndrome and dystonia. Am. *J. Med. Genet. Part A* **143**, 33–41
44. Shanske, S., Coku, J., Lu, J., Ganesh, J., Krishna, S., Tanji, K., Bonilla, E., Naini, A.B., Hirano, M. & DiMauro, S. (2008) The G13513A mutation in the *ND5* gene of mitochondrial DNA as a common cause of MELAS or Leigh syndrome: evidence from 12 cases. *Arch. Neurol.* **65**, 368–372
45. Valente, L., Piga, D., Lamantea, E., Carrara, F., Uziel, G., Cudia, P., Zani, A., Farina, L., Morandi, L., Mora, M. et al. (2009) Identification of novel mutations in five patients with mitochondrial encephalomyopathy. *Biochim. Biophys. Acta* **1787**, 491–501
46. Tawata, M., Hayashi, J.-I., Isobe, K., Ohkubo, E., Ohtaka, M., Chen, J., Aida, K. & Onaya, T. (2000) A new mitochondrial DNA mutation at 14577 T/C is probably a major pathogenic mutation for maternally inherited type 2 diabetes. *Diabetes* **49**, 1269–1272

Biochem. Soc. Symp. 78
Citation reference: Biochem. Soc. Trans. (2011) **39**, 807–811.

13

The choreography of multidrug export

Rupak Doshi, Daniel A.P. Gutmann, Yvonne S.K. Khoo, Lisa A. Fagg and Hendrik W. van Veen[1]

Department of Pharmacology, University of Cambridge, Tennis Court Road, Cambridge CB2 1PD, U.K.

Abstract

Multidrug transporters have a crucial role in causing the drug resistance that can arise in infectious micro-organisms and tumours. These integral membrane proteins mediate the export of a broad range of unrelated compounds from cells, including antibiotics and anticancer agents, thus reducing the concentration of these compounds to subtoxic levels in target cells. In spite of intensive research, it is not clear exactly how multidrug transporters work. The present review focuses on recent advancements in the biochemistry and structural biology of bacterial and human multidrug ABC (ATP-binding cassette) transporters. These advancements point to a common mechanism in which polyspecific drug-binding surfaces in the membrane domains are alternately exposed to the inside and outside surface of the membrane in response to the ATP-driven dimerization of nucleotide-binding domains and their dissociation following ATP hydrolysis.

[1]*To whom correspondence should be addressed (email hwv20@cam.ac.uk).*

Introduction

Multidrug ABC (ATP-binding cassette) transporters have been the subject of intense study because they have been linked to failure of antimicrobial and anticancer chemotherapy [1,2]. These membrane proteins are ubiquitously distributed across all five kingdoms of life [3] and share a common four-domain organization. They usually contain two MDs (membrane domains) that comprise a hetero- or homo-dimer of two six-TMH (transmembrane helix) protomers [4]; the MDs contain the translocation pathway for drugs. The transporter is completed by two cytoplasmic NBDs (nucleotide-binding domains) that together generate the ATP-binding and hydrolysis-dependent conformational changes required for drug export [4]. The prototype multidrug ABC transporter ABCB1 is an example of this architecture, and has all four domains fused into a single polypeptide [5].

ABCB1 was discovered in 1976 by Juliano and Ling [6] in plasma membranes of colchicine-resistant CHO (Chinese-hamster ovary) cells. Since then, ABCB1 and its homologues in bacteria and humans have been noted for their startling ability to bind and transport a wide variety of structurally unrelated compounds, including HIV anti-retroviral drugs [7–9], cancer chemotherapeutics and antibiotics [10]. Detailed understanding of this property requires knowledge of the molecular basis of polyspecificity and the translocation mechanism of drugs.

X-ray crystal structures

After much anticipation, the first reported full-length ABC exporter crystal structure was that of dimeric Sav1866, a bacterial and functional homologue of the human ABCB1 [11,12]. This structure presented a medium-resolution view of an ABC exporter with the two NBDs bound to two nucleotide molecules at the domain interface and the two MDs spread apart, forming two wing-shaped structures. Because the internal chamber between the MDs was exposed to the extracellular space, this conformation was termed 'outward-facing'. The overall structural arrangement of the NBDs and TMHs (in the MDs) was consistent with previous cross-linking and EM (electron microscopy) studies of ABCB1 [11]. However, the formation of an NBD dimer with ADP alone was a surprising observation, because previous literature suggested the requirement of the γ-phosphate moiety of ATP for efficient NBD dimerization [13]. The authors later reported a second high-resolution crystal structure in which they successfully exchanged the ADP molecules with those of AMP-PNP (adenosine 5′-[β,γ-imido]triphosphate), a non-hydrolysable ATP analogue [14]. A similar outward-facing conformation of AMP-PNP-bound Sav1866 confirmed its interpretation as an ATP-bound state.

Almost simultaneously, three distinct crystal structures of the bacterial multidrug/Lipid A ABC exporter MsbA were published by Chang and

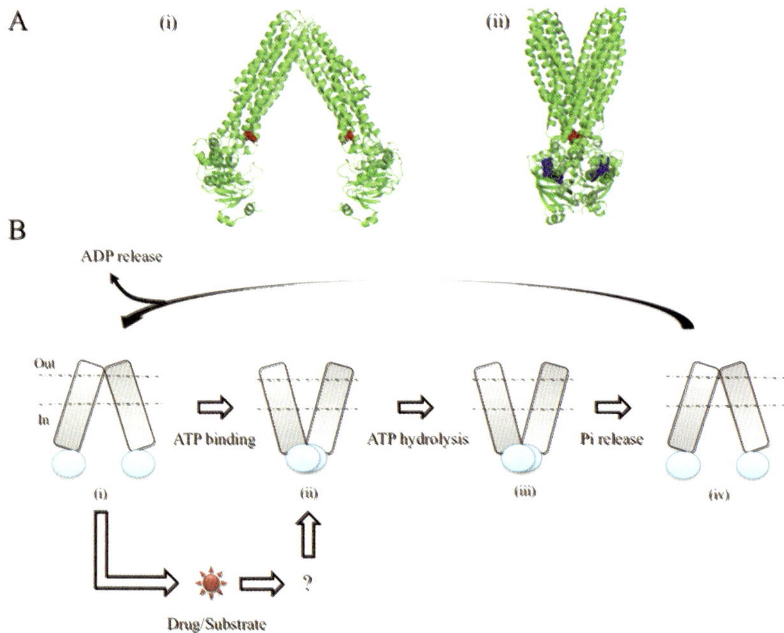

Figure 1. Conformational changes in MsbA
(**A**) The nucleotide-free inward-facing crystal structure (i) and nucleotide-bound outward-facing crystal structure (ii) of MsbA [15] are shown. AMP-PNP is depicted in blue. Glu^{208} (in red) was the target of an engineered cysteine residue in the cross-linking study [33] (see the main text for details). Co-ordinates of outward-facing MsbA (PDB code 3B6O) were obtained from the RCSB Protein Data Bank, whereas those for inward-facing MsbA were kindly provided by Professor Geoffrey Chang. (**B**) MsbA switches from inward-facing (i) to outward-facing (ii) on binding ATP, remains outward-facing while hydrolysing ATP to ADP · P_i (iii), and returns to inward-facing on the release of P_i (iv) [33]. The conformation of drug or Lipid A-bound MsbA and its influence(s) on the basal catalytic cycle are currently unknown. NBDs are depicted as blue circles, and MDs are depicted as grey bars. In and Out refer to the inside and outside of the cell respectively.

colleagues [15]. The first structure of MsbA (from *Escherichia coli*) is a nucleotide-free inward-facing conformation with the NBDs separated by a distance of ~50 Å (1 Å = 0.1 nm), in which the internal cavity at the MD interface faces the cell interior. This structure has an inverted V-shape (Figure 1A, i). The second structure of MsbA (from *Salmonella enterica* serotype Typhimurium) is an AMP-PNP-bound or ADP · V_i (V_i is inorganic vanadate)-trapped outward-facing conformation with the two wings as observed in the published structures of Sav1866 (Figure 1A, ii). The third structure of MsbA (from *Vibrio cholerae*) is a nucleotide-free inward-facing semi-closed conformation in which both the NBD and MD in one half-transporter are in close proximity to the corresponding domains in the other half-transporter, and in which the internal chamber is facing the cytosol.

The first mammalian multidrug ABC exporter was crystallized more recently [16]. This medium-resolution nucleotide-free structure of mouse ABCB1a was obtained in the inward-facing conformation and resembles the inverted V-shape structure of MsbA. Notably, the NBDs of ABCB1a are separated by ~30 Å, which is less than for inward-facing MsbA. The structure of ABCB1a shows a cytoplasmically open and hydrated cavity with a volume of approximately 6000 $Å^3$ at the interface between the two MDs. This chamber contains the bulk of residues previously identified as being important for drug binding (see below). Furthermore, the structure reveals a lateral portal between the two membrane halves that could serve as a pathway for drugs from the cytoplasmic membrane leaflet into the chamber. The same paper also presents structures of ABCB1 in complex with two distinct CCPIs (cyclic peptide inhibitors), one QZ59-RRR or two copies of QZ59-SSS, in the hydrophobic cavity. Binding of these two structurally distinct CCPIs in the same binding region is achieved by utilizing overlapping sets of non-polar phenylalanine and polar tyrosine residues belonging to TMHs 1, 6, 7 and 12. These account for 40% and 60% of interactions as measured by buried surface area respectively. The crystal structures also reveal that side-chain flexibility of these aromatic residues allows structural adjustments in the binding surface, thereby accommodating these sterically distinct, but equally hydrophobic compounds. Aliphatic residues in addition to hydrogen-bond donors and acceptors (serine, threonine and glycine) account for the remainder of interactions.

Biochemical and genetic studies of drug binding

The structural studies presented above are in agreement with biochemical observations. The importance of the membrane bilayer in drug binding and transport was established early on in the research of multidrug ABC transporters. For instance, the ability of anthracycline compounds to inhibit ABCB1-mediated extrusion of [^{3}H]daunorubicin is clearly correlated with their readiness to partition into the lipid bilayer [17], placing the site of drug binding in the MDs. Furthermore, FRET (fluorescence resonance energy transfer) measurements in ABCB1, between a covalently linked fluorescence acceptor (NBD-Cl) and bound fluorescent transport substrate Hoechst 33342 (donor), positioned drug binding close to the cystoplasmic leaflet of the membrane [18]. This notion was supported by fluorescence anisotropy and FRET experiments using the multidrug transporter substrate TMA-DPH [*N*,*N*,*N*-trimethyl-4-(6-phenyl-1,3,5-hexatrien-1-yl)phenylammonium *p*-toluenesulfonate]. In lactococcal cells, the rate of TMA-DPH extrusion by ABC and MFS multidrug transporters depended on the accumulation of the dye in the inner leaflet of the membrane [19]. Similar results were obtained more recently in human leukaemic cell lines expressing ABCB1 where drug transport was measured by following the fluorescent quenching TMA-DPH by the rhodamine analogue TMRM (tetramethylrhodamine methyl ester) [20]. Accumulation of TMRM in the cytoplasmic membrane leaflet was considerably higher in non-energized cells (ABCB1-inactive) or parental cell lines that did not express ABCB1 [20]. Along similar lines, the observation that the non-fluorescent precursors of calcein [calcein-AM (calcein acetoxymethyl ester)] and BCECF {BCECF-AM [2′,7′-bis-(2-carboxyethyl)-5(6)-carboxyfluorescein acetoxymethyl ester]} are extruded by ABCB1 and bacterial homologues before being converted into their fluorescent forms by cytoplasmic esterases [19,21] demonstrates that transport substrates can enter the binding site directly from the lipid phase. Equilibrium drug-binding studies, cysteine-scanning mutagenesis and photoaffinity labelling have provided useful insights into the approximate location and nature of drug binding within the protein (reviewed in [22]).

We examined published mutations in the human ABCB1 that alter multidrug transport in cells and identified mutations that affect the specificity for individual substrates (termed change-in-specificity mutations) [23]. When superimposed on the outward-facing crystal structure of homodimeric MsbA from *Salmonella* Typhimurium, these change-in-specificity mutations co-localize in a major groove in each of the two wings formed by the MDs [24]. Near the apex of the groove, TMH 6 in both monomers contains many change-in-specificity mutations and residues which, when replaced with cysteine residues in ABCB1, covalently interact with thiol-reactive drug analogues. We tested the importance of this region of TMH 6 for drug–protein interactions in *E. coli* MsbA. In particular, we focused on the conserved Ser^{289} and Ser^{290} in this region. Their simultaneous replacement with alanine (termed SASA mutant) significantly reduced the level of binding and transport of ethidium and anticancer drug Taxol by MsbA, whereas the interactions with Hoechst 33342 and the macrolide antibiotic erythromycin remained unaffected. Hence the SASA mutation is associated with a change-in-specificity phenotype analogous to that of the change-in-specificity mutations in ABCB1 [24]. This study is in agreement with the biochemical, genetic and structural data for ABCB1 [22], and suggests the presence of a drug-binding surface in the MD dimer, rather than drug-binding site, close to the leaflet/leaflet interface of the phospholipid bilayer.

Alternating access model

The transmembrane movement of substrates by transporters is based on a change in accessibility of substrate-binding sites between the inside surface and outside surface of the membrane [25]. In the ATP switch model, Higgins and Linton [26] proposed that ABC exporters switch from the inward-facing conformation to the outward-facing conformation by the ATP-binding-induced dimerization of the NBDs. This structural switch is thought to provide the so-called ‘power stroke’ for transport. Sauna and Ambudkar [27], however, stated that the ATP-induced dimerization of the NBDs alone is not sufficient to cause the high-to-low affinity transition because the occlusion (i.e. tight binding) of a nucleotide following the dimerization is required. Observations of nucleotide occlusion by ABCB1 during its catalytic cycle were also described by Tombline et al. [28].

In order to realistically envision the conformational changes occurring in a transporter while it progresses through its catalytic cycle, detailed structural information is required. For the last 4 years, various research groups have applied biochemical and biophysical techniques to MsbA to gain support for the alternating access model that was based on direct comparisons of the crystal snapshots of this protein. Having membrane proteins embedded in phospholipid bilayers is indispensible to the concept of alternating access. Mchaourab and co-workers made EPR and DEER (double electron–electron resonance) measurements on MsbA that was reconstituted into a membrane [29–31]. These elegant distance measurements confirmed a large NBD separation under nucleotide-free conditions, consistent with the inward-facing conformation, and NBD dimerization when trapped in the transition state with ADP · V_i. Cryo-EM studies undertaken on MsbA dimer particles in proteoliposomes found the transporter in an overall outward-facing conformation when bound to AMP-PNP or the transition state inducers ADP · V_i and ADP · AlF_3 [32]. These were the first results to provide strong biochemical and biophysical support for the published crystal structures.

ATPase cycle

In spite of the above-mentioned encouraging results, none of the X-ray crystallography, EPR, DEER or cryo-EM studies reported the conformation of MsbA at all four major steps of the catalytic cycle (namely, nucleotide-free, ATP-bound, ADP · P_i-bound, and ADP-bound and/or P_i-released). Thus a consolidated view of MsbA through an entire ATPase cycle remained to be achieved (for details, see Table 1 of [33]). Recently, however, this has been accomplished using disulfide cross-linking between the two halves of MsbA via cysteine residues (E208C–E208C′) engineered at the cytosolic extensions of the MDs [33]. To mimic the physiological setting as closely as possible, *E. coli* MsbA was analysed while present in a biological membrane (inside-out plasma membrane vesicles) and in reaction buffer and environmental conditions which support transport and ATP hydrolysis. These results strongly suggest that MsbA switches from a nucleotide-free inward-facing conformation to an outward-facing state on binding ATP. While remaining outward-facing on hydrolysing ATP to ADP · P_i, the MsbA transporter returns to the inward-facing resting state with ADP-bound, i.e. on the liberation of P_i (Figure 1B, i–iv). This represents the first structural description of the entire ATPase cycle of a single multidrug ABC exporter, using one and the same biochemical technique under near-physiological conditions.

Future directions in research

Our current insights into the structure and function of ABC exporters raise many new and interesting questions, some of which might be addressed in future work.

(i) Is ATP binding and hydrolysis the only form of metabolic energy driving transport by multidrug ABC exporters? Electrophysiological experiments and transport studies (using radioactive ions and fluorescent ion-selective probes) demonstrated that the bacterial ABCB1 homologue LmrA can transport cations (including organic molecules such as ethidium, and inorganic ions such as Na^+) by a secondary-active proton symport mechanism [34]. This work suggests that, although the overall transport mechanism of LmrA and other ABC systems is ATP binding/hydrolysis-dependent, certain conformational changes might be facilitated by electrochemical ion gradients.

(ii) How similar will bacterial and mammalian ABC exporter conformations be? The first crystal structures [16] and cryo-EM studies [35] on ABCB1 have revealed remarkable similarities to nucleotide-free inward-facing MsbA and nucleotide-bound outward-facing MsbA and Sav1866. However, more ligand-bound high-resolution structures, and biochemical and dynamic analyses based on EPR, DEER, cross-linking, FRET and others will be required to dissect the conformational cycle of mammalian ABC exporters.

(iii) Does drug/substrate binding alter the basal catalytic cycle? At present, the alternating access model suggests that ATP binding alone can initiate a catalytic cycle, even in the absence of added transport substrates. Thus, although this model explains how substrates may be exported, it fails to explain how the presence of drugs/substrates can stimulate the basal rate of ATP hydrolysis, which is an historical observation for most multidrug ABC exporters [36–38]. Drug binding-induced conformational changes in ABC exporters have been observed by various research groups [29,37,39–41], but the identification of this conformation and the inclusion of this conformation in the present scheme of alternating access is lacking.

(iv) Is it possible to use the knowledge of conformations to 'lock' multidrug exporters in an inhibited state? The ability to interrupt the activity of ABCB1 and others in tumours will require further optimization [42–44]. As our knowledge of the conformational dynamics of ABC exporters increases, it might be possible to design inhibitors that target specific conformations of the exporter, essentially 'locking' it down. Recently, mouse ABCB1a was co-crystallized with cyclic peptide inhibitors bound at the putative substrate-binding surfaces, yielding an inward-facing conformation [16]. Since these inhibitors were also shown to block the ATPase activity of ABCB1a, it is possible that the inhibition is based on the trapping of the transporter in the inward-facing conformation with disengaged NBDs, thus preventing NBD dimerization [16,22].

A comprehensive profile of the major conformations and drug-binding sites in ABCB1 and other multidrug transporters would be of great clinical and commercial benefit. For example, inhibition of the breast cancer resistance protein (ABCG2) in the gut by Elacridar has been found to greatly enhance the bioavailability of the topoisomerase inhibitor topotecan in anticancer treatment [45]. Hence newly developed inhibitors can be co-administered with chemotherapeutic drugs to enhance the pharmacokinetic properties of these drugs. This strategy will also be useful when aiming to reduce drug-efflux-based resistance in target cells. However, it should be noted that cells can activate various defence mechanisms when exposed to toxic drugs. If we aim to improve the efficiency of anticancer and antimicrobial chemotherapy through inhibition of drug efflux, it will be important to develop rapid diagnostic tools that identify those instances where drug efflux is the major contributor to drug resistance.

Acknowledgements

We apologize to all colleagues whose work was not cited in this brief review. H.W.v.V. thanks all present and past members of the laboratory for their contributions to the work described.

References

1. Garvey, M.I., Baylay, A.J., Wong, R.L. & Piddock, L.J. (2011) Overexpression of patA and patB, which encode ABC transporters, is associated with fluoroquinolone resistance in clinical isolates of *Streptococcus pneumoniae*. *Antimicrob. Agents Chemother.* **55**, 190–196
2. Pajic, M., Iyer, J.K., Kersbergen, A., Van Der Burg, E., Nygren, A.O., Jonkers, J., Borst, P. & Rottenberg, S. (2009) Moderate increase in Mdr1a/1b expression causes in vivo resistance to doxorubicin in a mouse model for hereditary breast cancer. *Cancer Res.* **69**, 6396–6404
3. Higgins, C.F. (1992) ABC transporters: from microorganisms to man. *Annu. Rev. Cell Biol.* **8**, 67–113
4. Saurin, W., Hofnung, M. & Dassa, E. (1999) Getting in or out: early segregation between importers and exporters in the evolution of ATP-binding cassette (ABC) transporters. *J. Mol. Evol.* **48**, 22–41
5. Ueda, K., Clark, D.P., Chen, C.J., Roninson, I.B., Gottesman, M.M. & Pastan, I. (1987) The human multidrug resistance (*mdr1*) gene: cDNA cloning and transcription initiation. *J. Biol. Chem.* **262**, 505–508
6. Juliano, R.L. & Ling, V. (1976) A surface glycoprotein modulating drug permeability in Chinese hamster ovary cell mutants. *Biochim. Biophys. Acta* **455**, 152–162
7. Hegedus, T., Orfi, L., Seprodi, A., Váradi, A., Sarkadi, B. & Kéri, G. (2002) Interaction of tyrosine kinase inhibitors with the human multidrug transporter proteins, MDR1 and MRP1. *Biochim. Biophys. Acta* **1587**, 318–325
8. Kim, R.B., Fromm, M.F., Wandel, C., Leake, B., Wood, A.J., Roden, D.M. & Wilkinson, G.R. (1998) The drug transporter P-glycoprotein limits oral absorption and brain entry of HIV-1 protease inhibitors. *J. Clin. Invest.* **101**, 289–294
9. Lee, C.G., Gottesman, M.M., Cardarelli, C.O., Ramachandra, M., Jeang, K.T., Ambudkar, S.V., Pastan, I. & Dey, S. (1998) HIV-1 protease inhibitors are substrates for the MDR1 multidrug transporter. *Biochemistry* **37**, 3594–3601
10. Higgins, C.F. (2007) Multiple molecular mechanisms for multidrug resistance transporters. *Nature* **446**, 749–757
11. Dawson, R.J. P. & Locher, K.P. (2006) Structure of a bacterial multidrug ABC transporter. *Nature* **443**, 180–185
12. Velamakanni, S., Yao, Y., Gutmann, D.A. & van Veen, H.W. (2008) Multidrug transport by the ABC transporter Sav1866 from *Staphylococcus aureus*. *Biochemistry* **47**, 9300–9308
13. Smith, P.C., Karpowich, N., Millen, L., Moody, J.E., Rosen, J., Thomas, P.J. & Hunt, J.F. (2002) ATP binding to the motor domain from an ABC transporter drives formation of a nucleotide sandwich dimer. *Mol. Cell* **10**, 139–149
14. Dawson, R.J. P. & Locher, K.P. (2007) Structure of the multidrug ABC transporter Sav1866 from *Staphylococcus aureus* in complex with AMP-PNP. *FEBS Lett.* **581**, 935–938

15. Ward, A., Reyes, C.L., Yu, J., Roth, C.B. & Chang, G. (2007) Flexibility in the ABC transporter MsbA: alternating access with a twist. *Proc. Natl. Acad. Sci. U.S.A.* **104**, 19005–19010
16. Aller, S.G., Yu, J., Ward, A., Weng, Y., Chittaboina, S., Zhuo, R., Harrell, P.M., Trinh, Y.T., Zhang, Q., Urbatsch, I.L. & Chang, G. (2009) Structure of P-glycoprotein reveals a molecular basis for polyspecific drug binding. *Science* **323**, 1718–1722
17. Friche, E., Demant, E.J., Sehested, M. & Nissen, N.I. (1993) Effect of anthracycline analogs on photolabelling of p-glycoprotein by [^{125}I]iodomycin and [^{3}H]azidopine: relation to lipophilicity and inhibition of daunorubicin transport in multidrug resistant cells. *Br. J. Cancer* **67**, 226–231
18. Qu, Q. & Sharom, F.J. (2002) Proximity of bound Hoechst 33342 to the ATPase catalytic sites places the drug binding site of P-glycoprotein within the cytoplasmic membrane leaflet. *Biochemistry* **41**, 4744–4752
19. Bolhuis, H., van Veen, H.W., Molenaar, D., Poolman, B., Driessen, A.J. & Konings, W.N. (1996) Multidrug resistance in *Lactococcus lactis*: evidence for ATP-dependent drug extrusion from the inner leaflet of the cytoplasmic membrane. *EMBO J.* **15**, 4239–4245
20. Katzir, H., Yeheskely-Hayon, D., Regev, R. & Eytan, G.D. (2010) Role of the plasma membrane leaflets in drug uptake and multidrug resistance. *FEBS J.* **277**, 1234–1244
21. Homolya, L., Holló, Z., Germann, U.A., Pastan, I., Gottesman, M.M. & Sarkadi, B. (1993) Fluorescent cellular indicators are extruded by the multidrug resistance protein. *J. Biol. Chem.* **268**, 21493–21496
22. Gutmann, D.A., Ward, A., Urbatsch, I.L., Chang, G. & van Veen, H.W. (2010) Understanding polyspecificity of multidrug ABC transporters: closing in on the gaps in ABCB1. *Trends Biochem. Sci.* **35**, 36–42
23. Shilling, R.A., Venter, H., Velamakanni, S., Bapna, A., Woebking, B., Shahi, S. & van Veen, H.W. (2006) New light on multidrug binding by an ATP-binding-cassette transporter. *Trends Pharmacol. Sci.* **27**, 195–203
24. Woebking, B., Velamakanni, S., Federici, L., Seeger, M.A., Murakami, S. & van Veen, H.W. (2008) Functional role of transmembrane helix 6 in drug binding and transport by the ABC transporter MsbA. *Biochemistry* **47**, 10904–10914
25. Seeger, M.A. & van Veen, H.W. (2009) Molecular basis of multidrug transport by ABC transporters. *Biochim. Biophys. Acta* **1794**, 725–737
26. Higgins, C.F. & Linton, K.J. (2004) The ATP switch model for ABC transporters. *Nat. Struct. Mol. Biol.* **11**, 918–926
27. Sauna, Z.E. & Ambudkar, S.V. (2007) About a switch: how P-glycoprotein (ABCB1) harnesses the energy of ATP binding and hydrolysis to do mechanical work. *Mol. Cancer Ther.* **6**, 13–23
28. Tombline, G., Muharemagic, A., White, L.B. & Senior, A.E. (2005) Involvement of the occluded nucleotide conformation of P-glycoprotein in the catalytic pathway. *Biochemistry* **44**, 12879–12886
29. Borbat, P.P., Surendhran, K., Bortolus, M., Zou, P., Freed, J.H. & Mchaourab, H.S. (2007) Conformational motion of the ABC transporter MsbA induced by ATP hydrolysis. *PLoS Biol.* **5**, e271
30. Zou, P., Bortolus, M. & Mchaourab, H.S. (2009) Conformational cycle of the ABC transporter MsbA in liposomes: detailed analysis using double electron–electron resonance spectroscopy. *J. Mol. Biol.* **393**, 586–597
31. Zou, P. & Mchaourab, H.S. (2009) Alternating access of the putative substrate-binding chamber in the ABC transporter MsbA. *J. Mol. Biol.* **393**, 574–585
32. Ward, A., Mulligan, S., Carragher, B., Chang, G. & Milligan, R.A. (2009) Nucleotide dependent packing differences in helical crystals of the ABC transporter MsbA. *J. Struct. Biol.* **165**, 169–175
33. Doshi, R., Woebking, B. & van Veen, H.W. (2010) Dissection of the conformational cycle of the multidrug/lipid A ABC exporter MsbA. *Proteins* **78**, 2867–2872
34. Velamakanni, S., Lau, C.H.F., Gutmann, D.A.P., Venter, H., Barrera, N.P., Seeger, S.A., Woebking, B., Matak-Vinkovic, D., Balakrishnan, L., Yao, Y. et al. (2009) A multidrug transporter with a taste for salt. *PLoS ONE* **4**, e6137

35. Lee, J.-Y., Urbatsch, I.L., Senior, A.E. & Wilkens, S. (2008) Nucleotide-induced structural changes in P-glycoprotein observed by electron microscopy. *J. Biol. Chem.* **283**, 5769–5779
36. Eckford, P.D.W. & Sharom, F.J. (2008) Functional characterization of *Escherichia coli* MsbA. *J. Biol. Chem.* **283**, 12840–12850
37. Szabó, K., Welker, E., Bakos, É., Müller, M., Roninson, I., Váradi, A. & Sarkadi, B. (1998) Drug-stimulated nucleotide trapping in the human multidrug transporter MDR1. *J. Biol. Chem.* **273**, 10132–10138
38. Scarborough, G. (1995) Drug-stimulated ATPase activity of the human P-glycoprotein. *J. Bioenerg. Biomembr.* **27**, 37–41
39. Siarheyeva, A. & Sharom, F.J. (2009) The ABC transporter MsbA interacts with lipid A and amphipathic drugs at different sites. *Biochem. J.* **419**, 317–328
40. Loo, T.W., Bartlett, M.C. & Clarke, D.M. (2003) Substrate-induced conformational changes in the transmembrane segments of human P-glycoprotein. *J. Biol. Chem.* **278**, 13603–13606
41. Liu, R. & Sharom, F.J. (1996) Site-directed fluorescence labeling of P-glycoprotein on cysteine residues in the nucleotide binding domains. *Biochemistry* **35**, 11865–11873
42. Bonate, P.L., Howard, D.R. & Thompson, T.N. (2011) The clinical significance of drug transporters in drug disposition and drug interactions. *Pharmacokinetics in Drug Development*, Springer, New York 285–313
43. Fletcher, J.I., Haber, M., Henderson, M.J. & Norris, M.D. (2010) ABC transporters in cancer: more than just drug efflux pumps. *Nat. Rev. Cancer* **10**, 147–156
44. Szakács, G., Paterson, J.K., Ludwig, J.A., Booth-Genthe, C. & Gottesman, M.M. (2006) Targeting multidrug resistance in cancer. *Nat. Rev. Drug Discovery* **5**, 219–234
45. Kuppens, I.E., Witteveen, E.O., Jewell, R.C., Radema, S.A., Paul, E.M., Mangum, S.G., Beijnen, J.H., Voest, E.E. & Schellens, J.H. (2007) A phase I, randomized, open-label, parallel-cohort, dose-finding study of elacridar (GF120918) and oral topotecan in cancer patients. *Clin. Cancer Res.* **13**, 3276–3285

Keyword index